Particle Technology Forum 2021

Held at the 2021 AIChE Annual Meeting

Boston, Massachusetts, USA and Online
7 - 11 November and 15 - 19 November 2021

ISBN: 978-1-7138-5699-3

Printed from e-media with permission by:

Curran Associates, Inc.
57 Morehouse Lane
Red Hook, NY 12571

AIChE
120 Wall Street, FL 23
New York, NY 10005-4020

Phone: (800) 242-4363
Fax: (203) 775-5177

www.aiche.org

Additional copies of this publication are available from:

Curran Associates, Inc.
57 Morehouse Lane
Red Hook, NY 12571 USA
Phone: 845-758-0400
Fax: 845-758-2633
Email: curran@proceedings.com
Web: www.proceedings.com

TABLE OF CONTENTS

630151 Influence of Interfaces in Electrical Properties of 3D Printed Structures

Abstract:

Additive Manufacturing technology has advanced to being a major player in the field of manufacturing in the last two decades revolutionizing the manufacturing of mechanical parts for aerospace and automotive applications, prosthetics, and dentistry. The application of additive manufacturing in the production of sensors and actuators can alleviate the delivery and storage related constraints for remote locations such as offshore platforms and space stations. Material extrusion-based additive manufacturing (MEAM) is a uniquely suited additive manufacturing technique for cost-effective printing of sensors and actuators. In MEAM, a 3D structure is built in a layer-by-layer manner as instructed by a g-code, by extruding a thermoplastic material through a nozzle heated above the melting point of the plastic. Several parameters including extrusion temperature, print speed, bed temperature, part orientation, layer height, raster angle, and infill can be varied in MEAM. Contemporary simulation and experimental studies, both show that the print parameters play an important role in determining the strength of interlayer bonding and the resulting anisotropy in mechanical strength of printed parts. To develop MEAM into an industrially feasible technique for 3D printing sensor and actuator components, similar fundamental studies are necessary to predict electrical part performance based on print parameters.

In this study, we conduct work using a commercial conductive composite of polylactic acid and carbon black. We adopt a hollow box structure specimen to examine the effect of extrusion temperature and print speed on the electrical characteristics of the bond interfaces, specifically those between two subsequent layers. The walls of the hollow box compose of single extruded fibers stacked on top of each other. The electrical characteristics are measured using small signal impedance spectroscopy to avoid Joule heating effects. The impact of varying the extrusion temperature and print speed on electrical impedance was examined in two directions, across the interlayer bond interfaces (Z-direction), and along the interlayer bond interfaces (F-direction). A variation of impedance in F-direction with extrusion temperature was observed and not print speed, indicating that the material resistivity varied with extrusion temperature. Thus, the Z/F ratio was used to quantify anisotropy in electrical impedance as the extrusion temperature and print speed were varied. It was found that the Z/F ratio remains relatively constant with variation in extrusion temperature and print speed. Variation in sample cross-section in Z- and F- direction were found to result in a Z/F ratio of 1 in COMSOL Multiphysics simulations. This showed that the bond interfaces were primarily responsible for the observed Z/F of 2.15 ± 0.23, that is, the bond interfaces contribute nearly two-third of the impedance when measuring across the fibers.

To understand how the impedance scales with the number of interfaces, impedance was measured across different number of bond interfaces in samples printed with extreme parameters - high/low print speed and high/low extrusion temperature. Impedance was found to scale linearly with the number of interfaces. This implied that tests across single bond interfaces could be used to predict results over multiple interfaces; thus, allowing to study effects of multiple parameters in a relatively short time. The scaling law was found to change when using different extrusion temperature; high extrusion temperature has a lower impedance per interface.

Thus by investigating the electrical behavior of bond interfaces, we provide a framework to model and predict anisotropy in electrical impedance and predict electrical impedance based on print parameters.

630223 Modeling Heat Transfer in Material Extrusion Additive Manufacturing: Balancing Model Resolution with Computational Demands

Abstract:

Thermal modeling informs the understanding of cooling phenomena and structural evolution during material extrusion (MatEx) additive manufacturing to help predict and tune final part properties. For fused filament fabrication (FFF) and big area additive manufacturing (BAAM), in particular, thermal models are well-suited to augment parameter optimization and closed-loop control for high throughput manufacturing of custom parts and products. However, limitations in computational time, speed, and cost pose a challenge to modeling large structures with sufficient multi-dimensional granularity over extended timescales, with many models confined either to high resolution studies on short length and timescales or low-resolution, lumped approximations for larger simulations. Thus, the goal of this work is to develop a high fidelity thermal model applicable from desktop to BAAM systems with reduced computational constraint.

We present a scalable, two-dimensional finite volume model and algorithm to rapidly simulate thermal profiles across both FFF and BAAM length and time scales. Our approach implements a two-dimensional implicit numerical strategy that confers reduced computational complexity while offering fine scale granularity in capturing critical inter- and intralayer cooling dynamics and radiative heat transfer. Further, analysis of the effective Biot number (Bi) reveals that, on the BAAM scale, layer width and radiative heat transfer both drive the formation of intralayer temperature gradients, exceeding 51°C between the layer center and edge. We further simulate the fabrication of single road-width acrylonitrile-butadiene-styrene (ABS) walls at the BAAM scale (50 layers) to demonstrate the relative speed of the computational approach and its ability to capture the complex dynamics of a multilayer build. The model agrees with published experimental measurements, and results gauge the influence of layer thickness and proximity to print bed on cooling behavior, with larger layers and separation from the print bed lending to longer periods for interlayer welds to form above the ABS glass transition temperature (T_g). Finally, a BAAM case study highlights the model's utility in design for additive manufacturing (DfAM) and print optimization, demonstrating how changes in bed temperature and layer time can impact interlayer bonding.

631340 Assessing the Fidelity of Additively Manufactured Objects

Abstract:

Despite the rapid growth of additive manufacturing (AM), its broad implementation is still hindered by challenges including limited material options, inconsistent part quality, and fabrication defects. Although current additive manufacturing technologies use high-precision machines, in practice, manufactured parts can deviate significantly in both shape and structure. Thus, characterization of the fidelity of manufactured objects is critical to establish and ensure performance. Although there are many ways to compare printed objects and the digital counterparts from which they are created, many of these methods are biased due to the way that measurements are made or the subjectivity of alignment procedures. To remove these subjectivities and biases, alternative printability metrics were developed based on a new alignment technique. Rather than using subjectively selected landmarks, the new method uses the axis of the minimum moment of inertia as reference. A boundary-based printability index (PI_B) and a volume-based printability index (PI_v) were then used to assess external dimensions and object volume. X-ray computed tomography (X-CT) images of 3D printed objects were used. Guidelines for statistically sound sampling were also developed. The results show the extent to which different metrics are sensitivity to different types of flaws and offer ways of comparing the effect of processing conditions on print fidelity.

631938 Design and Processing of Open Lattice Structures for Tunable Fluid Phenomena

Abstract:

The rapid evolution of additive manufacturing has realized previously unproducible designs, such as intricate lattice networks, which have demonstrated utility in heat exchangers, reaction vessels, structural scaffolds, biomedical implants, and even consumer products. A growing body of research has considered this class of materials for applications in process intensification, thanks to the variety of available designs and the range of available dimensions. However, much of the basis for these investigations is founded on historical use of stochastic open cell foams, and thus the potential of additive manufacturing for designing purposely tailored structures for fluid applications remains largely untapped. Furthermore, process variability further contributes hindrances to the adoption of these technologies. Here we investigate an assortment of lattice architectures in the context of fluid dynamics applications. Using the production-oriented Carbon M1 3D printer, we produce lattice parts for experimental pressure studies with homogeneous and heterogeneous unit cell configurations. Upon dimensional characterization, we observe that the photopolymerization process results in a scale-dependent gradient deviation from as-designed part dimensions, to which fluid processes may be more sensitive than structural applications. In response, we demonstrate a process for design-stage compensation to improve agreement between model predictions and experimental results. We further examine how scaling of the unit cell and overall part geometry contribute to fluid phenomena within these ordered open cellular structures and present guidelines for designing lattice configurations. With this complete pipeline, we illustrate the potential for additive manufacturing to produce tailor-made, multifunctional process components for chemical engineering applications.

621938 Methods to Evaluate Residual Stress in FDM Printed Parts

Abstract:

Additive manufacturing has gained a lot of attention lately because of its ability to develop prototypes directly from a CAD model. However, due to the layer-by-layer sequential solidification of the printed material and large temperature differences that result during the printing process, residual stresses are prone to develop in AM parts, leading to geometrical inaccuracies like volume shrinkage, warpage and distortion of the printed part, adversely affecting the final product quality. Measuring and controlling such residual stresses can therefore help us better choose the printing parameters and hence greatly enhance the dimensional accuracy of the printed parts. In this study, Fused deposition Modeling (FDM) is employed to study the effect of printing speed, printing temperature and nozzle diameter on the residual stress generated in commonly used FDM filaments like ABS and PC printed parts. Volume shrinkage of ABS printed samples is quantified suing Dynamic Mechanical Analysis (DMA). The destructive hole-drilling technique coupled with a 2D subset-based digital image correlation (DIC) software is used to measure the residual strains. Non-destructive optical residual stress measuring techniques like RAMAN spectroscopy and birefringence are also used for characterization. Finally, an experimentally validated finite element analysis (FEA) model is developed to simulate the FDM printing process and the hole-drilling method. A non-linear temperature-dependent viscoelasticity model is also developed to predict the behavior change of the filament as it solidifies. The simulations and models developed can be directly applied to predict the residual stresses that are generated during the FDM printing process and hence fine tune the printing parameters leading to an improved final product quality.

622870 Characterizing and Modeling Structured Bubbling in Gas-Solid Fluidized Beds

Abstract:

Bubble dynamics have a strong effect on particle conveyance and mixing as well as gas-solid contact in fluidized beds. Bubbles typically rise and coalesce in a chaotic manner in fluidized beds, making industrial systems different to scale up and operate in an optimal manner. Prior studies have shown that novel device designs, such as oscillating gas flow rate, can induce structure in fluidized beds. Here, we discuss our ongoing research of structured bubble dynamics in fluidized beds. This includes (i) physical mechanisms which create structure, (ii) quantification of structuring, (iii) analytical modeling and design rules for resonance in fluidized beds, (iv) effect of scale-up on structure and (v) development of continuum computational modeling which can capture structured bubbling dynamics.

634364 Modeling of Taylor-Couette Vortices in Fluidized Beds

Abstract:

Despite the experimental discovery of Taylor-Couette vortices emerging in fluidized beds subject to shear in a Couette cell in 2004 (Conway et al., *Nature* 2004), the physics and potential applications of this phenomenon have yet to be investigated significantly since. Here, we utilize computational modeling to reproduce the phenomenon, investigate its origin and regimes of behavior.

623908 Fluid-Mediated Sources to Granular Temperature in Homogeneous Fluidization

Abstract:

When considering the dynamics of rapid particle-laden flows, a hydrodynamic description for the solids phase is often adopted. Drawing from a strong analogy with molecular fluids, Enskog kinetic theory has served as a valuable starting point for the modelling of particle-laden suspensions. Traditionally, one considers the classic Chapman-Enskog expansion of the Boltzmann equation to arrive at a Navier-Stokes equation with detailed expressions for the transport coefficients. This approach encompasses the classic two-fluid model. In many cases, the role of a surrounding viscous fluid is neglected. However, it has become increasingly clear that fluid ow through an assembly of particles leads to a distribution of hydrodynamic forces with significant variance. Variance in drag force, arising from neighbor-induced perturbations to the flow field, acts as sources and sinks to the granular temperature balance. Here, we describe drag force disturbances as a stochastic quantity that follows an Ornstein-Uhlenbeck process. For the case of homogeneous fluidization, we derive the sources and sinks to granular temperature by analytically solving the Fokker-Planck equation in force-velocity phase space and integrating over the relevant quadrants. The obtained solutions may be utilized for comparison with PR-DNS benchmark data as well as employed within the classic two-fluid model.

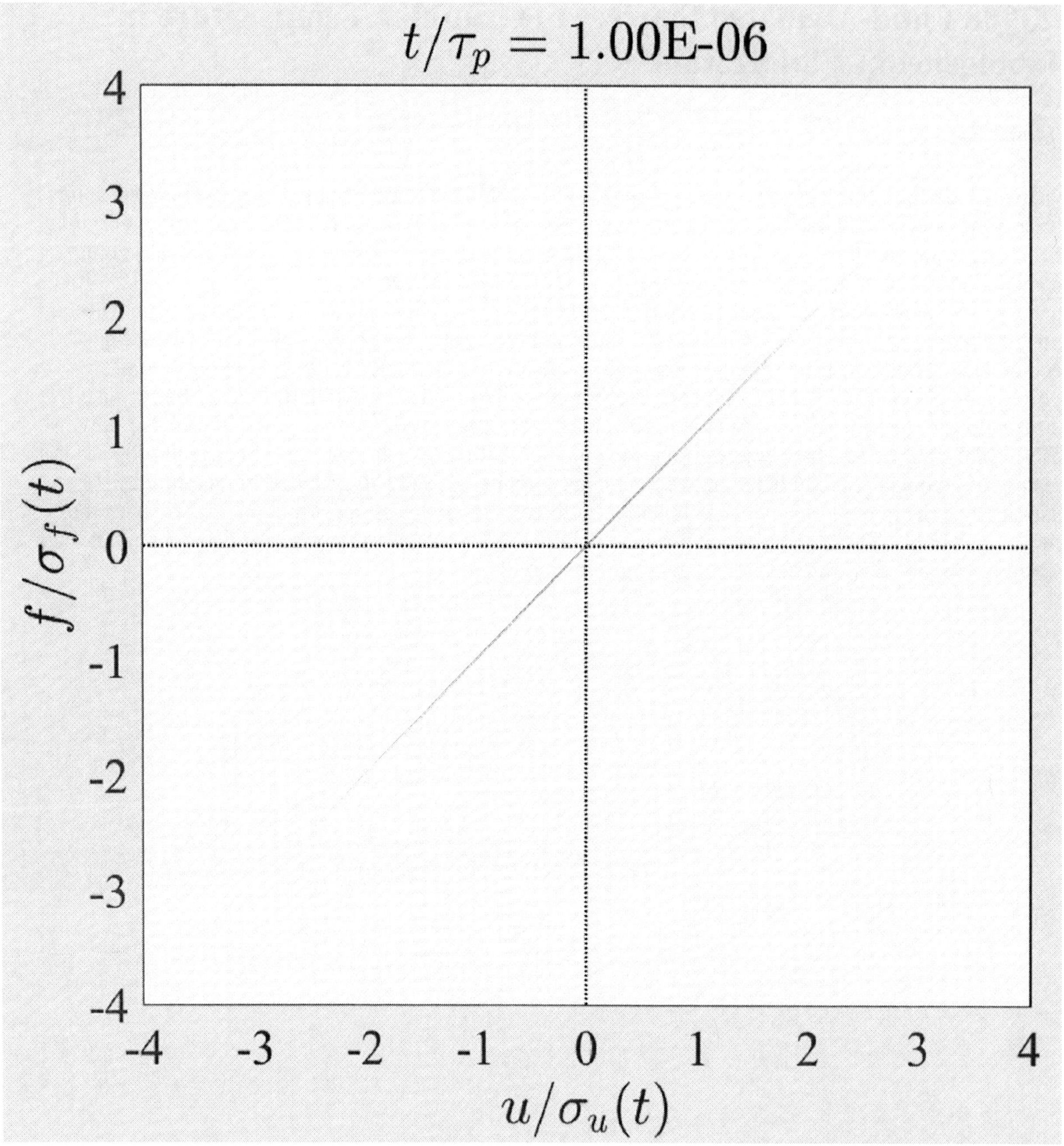
$t/\tau_p = 1.00E\text{-}06$
4
3
2
1
0
-1
-2
-3
-4
$f/\sigma_f(t)$
-4
-3
-2
-1
0
1
2
3
4
$u/\sigma_u(t)$

624328 Effect of Particle Size on the Modulation of Near-Wall Turbulent Flow Structures in Particle-Resolved Direct Numerical Simulations and Eulerian-Lagrangian Simulations

Abstract:

The modification of turbulent structures by finite-size and point particles suspended in a fluid flow is of interest to pipeline design for oil and slurry transport. Particle interactions with turbulent flow structures is investigated in a turbulent channel flow at a friction Reynolds number of 180, particle volume fraction 0.01, and particle-fluid density ratio 2.6. To understand the role of particle size on the modulation of flow structures, two computational approaches are deployed. Simulations with finite-size particles use fully resolved particles, referred to as Particle-Resolved DNS (PR-DNS). Simulations with point particles use a Eulerian-Lagrangian strategy (EL). The PR-DNS is enabled by a novel immersed boundary (IB) method derived from the volume-filtering framework. In this method, the fluid-particle coupling is obtained by direct integration of fluid stresses on the surface of particles. These integrals are discretized using a tessellation of the surfaces, and used to update the particle position and rotation through the fluid. Conservation equations for the fluid phase are solved on a Cartesian grid. The numerical approach is second order accurate, robust and stable for simulations with arbitrarily-shaped moving particles. We show that this PR-DNS strategy is in good agreement with experimental data for a sedimenting particle in a viscous fluid. The EL simulation strategy is also based on the volume-filtering method and has been previously thoroughly tested in massively parallel simulations of particle-laden flows. In the PR-DNS, we consider finite-size particles with diameter equal to 16 in wall units, leading to a total of 16,796 fully resolved particles. In the EL simulations, the point particles have a diameter equal to 0.5 in wall units, resulting in 134 million particles in total. Additional statistics are collected from an auxiliary simulation with no particles and are compared to the statistics found in the particle-laden turbulent channel flow simulations. To reveal the qualitatively different modulation mechanisms, we analyze the effect of point and finite-size particles on the carrier fluid statistics and near-wall coherent structures. Further, the clustering patterns between EL simulations and PR-DNS are compared. The simulations reveal two distinct flow modulation mechanisms. Point particles act collectively to modulate the carrier flow through a vortex-dissipating effect, whereas finite-size particles have a vortex-shredding effect caused by direct collisions between individual particles and near-wall coherent structures.

639587 Particulate Process and Product Design: Single Drop Granulation

Abstract:

Many industries, including pharmaceuticals, food, agricultural chemicals, detergents, catalysts, and consumer products, manufacture a variety of products from particulate materials. Product or intermediate specifications may include size, size distribution, shape, strength, porosity, and content uniformity of particles or of particulate-based delivery forms. The aim of this work is to design particulate products and processes through the study of fundamental phenomena involved in particle-particle, particle-fluid, and particle-equipment interactions to produce required product specifications. By working backwards from the desired end product attributes, this multiscale design approach covers the entire sequence of raw feed formulation handling through equipment processing to the final granular product. This talk will highlight single drop granulation as a form of particulate process and product design.

621066 A Stimuli-Responsive, Hydrolysable Poly(Vinyl Laurate-co-Vinyl Acetate) Nanoparticle Platform for in Situ Release of Surfactants

Abstract:

Efficient surfactant delivery with controlled release is of great practical interest in a number of fields including oil spill remediation and oil recovery.[1] However, premature adsorption of the surfactant at various surfaces and interfaces reduces their efficiency and limits practical applications. One proposed approach to address this challenge is to encapsulate the surfactant into an appropriate host. In most systems, the immobilization relies on relatively weak interactions between the surfactant molecules and the host, which leads to premature release of the surfactant.[2,3] A stimuli-responsive, sub-100nm nanoparticle platform with a hydrolysable ester side chain for in-situ generation of surfactants is demonstrated (Scheme. 1). The nanoparticles serve as a delivery vehicle as well as a precursor for the surfactant, where the slow and controlled hydrolysis of the ester groups generates carboxylic acid, which subsequently reacts with the cations in the medium to produce the surfactant. The nanoparticles were synthesized via copolymerization of vinyl-laurate and vinyl-acetate (P(VL-co-VA), 3:1 molar ratio) and stabilized with a protective poly(ethylene-glycol) (PEG) shell. The NPs are ~ 55 nm in diameter with a zeta potential of -54 mV. Hydrolysis kinetics in an accelerated, base-catalysed reaction show release of about 11 and 31% of available surfactant at 25 and 80°C, respectively. The corresponding values in saline water are 22 and 76%. The efficiency of the released surfactant in reducing the interfacial tension (IFT), altering wettability, and stabilizing an oil-water emulsion was investigated through contact angle measurements and laser confocal scanning microscopy (LCSM) and benchmarked to sodium laurate (SL), a commercially available surfactant representing the product derived upon hydrolysis of the nanoparticles. All these measurements demonstrate both the efficacy of the NP system for surfactant delivery and the ability of the released surfactant to alter wettability and stabilize an oil-water emulsion.

References:

(1)Farinmade, A. et al. *ACS Appl. Mater. Interfaces* **2020**.

(2)de Freitas, F. et al. *Fuel* **2019**, *241*.

(3)De Avila, J. et al. *J. Appl. Polym. Sci.* **2016**.

Scheme 1 Hydrolysis scheme of P(VL-co-VA) NPs and application in wettability alteration and stabilization of emulsions

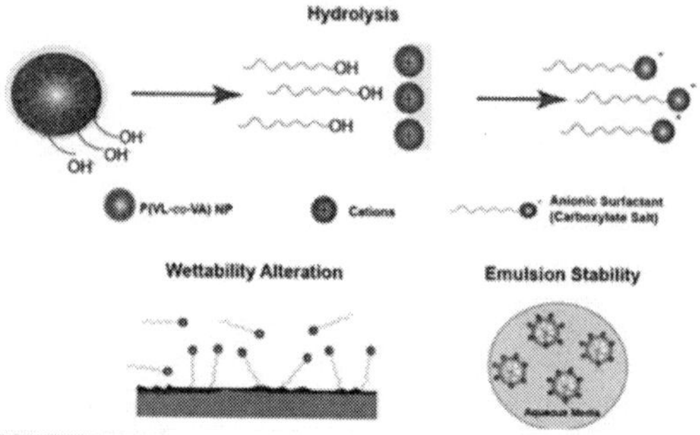

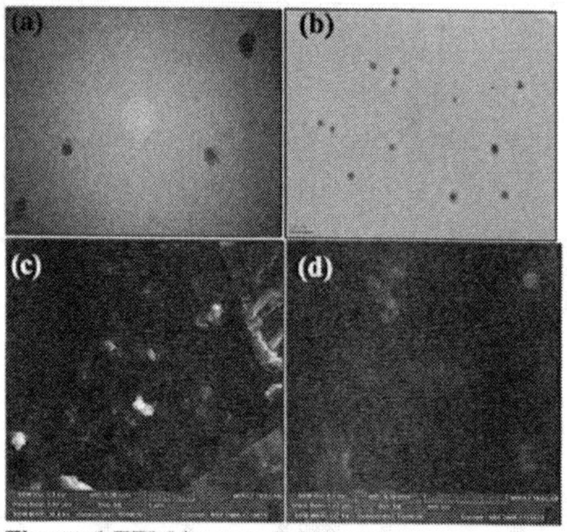

Figure 1 TEM images (a&b) and corresponding SEM images (c&d) of *unhydrolyzed* P(VL-co-VA) NPs.

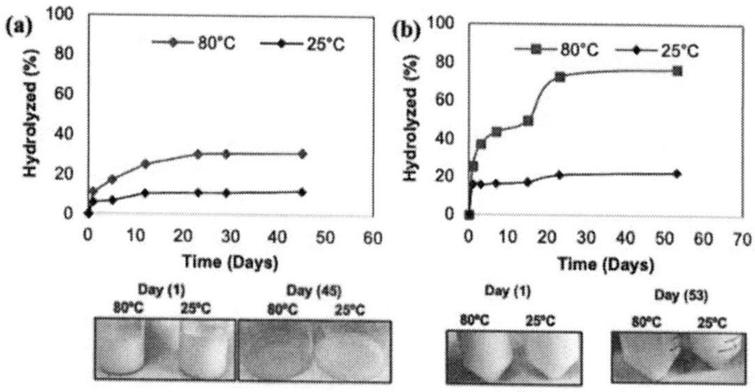

Figure 2 (a) Hydrolysis profile of P(VL-co-VA) NPs in 0.1M NaOH at 80°C and 25°C dispersed in (a) deionized water, (b) high salinity water.

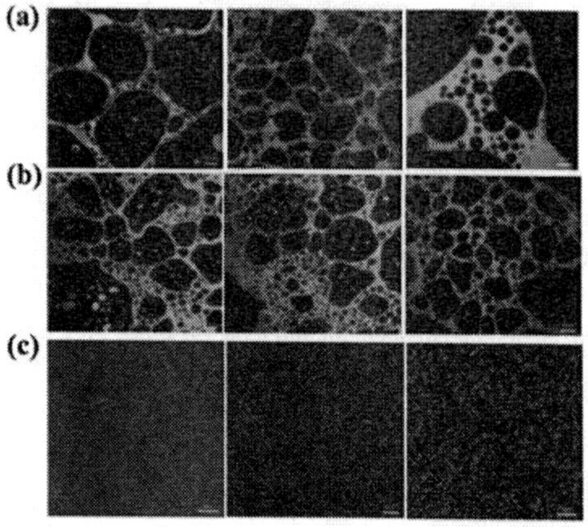

Figure 3 Confocal Micrographs of (a) crude oil in deionized water, DI, (b) unhydrolyzed P(VL-co-VA) NPs in crude oil/DI water, (c) hydrolyzed P(VL-co-VA) NPs in crude oil/DI water. The volume ratio of oil to water is 1:1.

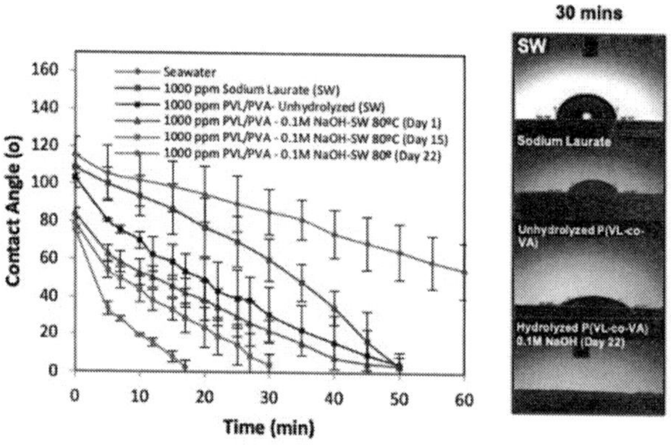

Figure 4 Contact angle profiles and corresponding optical images taken on a calcite pellet saturated with model oil (0.01M stearic acid in hexadecane). Measurements were conducted in seawater. The profiles are for droplets consisting of water, 1000 ppm sodium laurate surfactant, 1000 ppm unhydrolyzed P(VL-co-VA) nanoparticles and the hydrolyzed P(VL-co-VA) in NaOH at 80°C.

630097 Surface Characterization of Modified Fe_3O_4 Catalysts for Inductively Driven Alcohol Oxidation

Abstract:

Magnetic nanoparticles (NPs) such as iron oxide (Fe_3O_4) generate local heating under an alternating magnetic field (AMF), and they have been used for magnetic hyperthermia, drug delivery, and catalysis. Induction heating (IH) catalysis has shown to be a prominent field due to the potential decrease in heat dissipation losses. Since the catalyst is also the heat source, energy transfer to reactants is facilitated. Optimum Fe_3O_4 nanocatalysts are generally synthesized with the aid of surfactants, which allows tuning of catalyst morphology and size (surface area). However, the application of such magnetic catalysts presents a few challenges, namely 1) accurate determination of the surface temperature (i.e., reaction temperature) and 2) efficient surfactant removal that hinders reactant accessibility and heat transfer, 3) catalytic deactivation due to coking/oxidation of Fe_3O_4. Despite advances in the development of thermal probes that are not susceptible to an AMF, it is still challenging to quantify the NP's surface temperature instead of relying on bulk measurements to quantify heat transfer, which is key in catalysis. This work uses a photoluminescent (PL) shell to determine the catalyst's surface temperature in situ using europium doped yttrium vanadium oxide (YVO_4:Eu^{3+}). By incorporating this shell on the NPs, the surface temperature can be quantified based on the change in PL intensity with temperature. Afterward, the Fe_3O_4 surface temperature can be calibrated to a set AMF to study the role of IH on the condensation of 1-octanol. The catalyst is then modified with respect to its surface chemistry, morphology, and doping concentration ($Fe_{3-x}Cr_xO_4$) to evaluate the effects on reaction selectivity and yield.

This work demonstrates that the reaction of 1-octanol on Fe_3O_4 structures can be tuned to increase aldehyde yield by 2.5-fold via modification of the surface ligands and by performing the reaction inductively. Notably, product selectivity also changes to favor the production of the octanol ester once highly faceted and asymmetric star-shaped nanoparticles are used. Additionally, the catalyst's redox properties can be tuned via Cr doping, modifying the IH profile and yields/activities. These studies are facilitated by the determination of the surface temperature, where the sample is placed in an AMF combined with a PL spectrometer using fiber-optic light guides. The change in PL intensity with an applied field is compared to a similar analysis using a controlled temperature stage. The surface temperatures are up to 65 °C higher than bulk, requiring adjustment of the AMF for IH catalysis at the desired reaction temperature (200 °C). The core NPs are surface treated to allow accessibility of the reactants, and interestingly, this treatment also enhances heat transfer. As a comparison, surfactant-less NPs are also tested but are outperformed by phase-transferred colloidal NPs. The product yield with treated colloidal NPs increases by 6-fold (IH) and 2.5-fold (thermal heating) compared to surfactant-less. These results demonstrate the advantages of using IH technology for efficient energy consumption. Furthermore, the incorporation of Cr has the potential to stabilize Fe^{+2} (Fenton active species), expanding the applications of this catalyst to other reactions and potentially decreasing the rate of Fe^{2+} oxidation to weakly magnetic and lower IH temperature catalysts, which improves its lifetime.

631123 Heat Transfer in a Rotary Drum

Abstract:

Rotary drums deliver high heat and mass transfer rates, making them extensively useful for powder mixing and heating processes in metallurgical, cement, mining, pharmaceutical, detergent and other particulate processing applications. However, these processes are very difficult to model and operate because of the complex behavior affected by the particle parameters and process conditions such as particle size and distribution, shape, composition, fill level, rotation rate, and operating temperature. The current experimental research is focused on understand the heating mechanisms by conduction and radiation at high temperatures (>600ºC) and studying the effects of particle size distribution, fill level, and relatively low rotation rate (<10rpm). This work also serves to bridge the gap in the literature for experimental data and validating simulation-based research.

Experiments are performed using 2 mm diameter silica beads to investigate the granular flow and heat transfer by conduction and radiation inside a 6-inch diameter and 3-inch-long stainless-steel rotary drum. The drum is attached to two 10-inch titanium wheels, hence preventing the direct contact of the drum wall with the rollers. One end of the drum is closed using a transparent quartz window, capable of handling high temperatures. Another end of the drum is closed using a sapphire window, compatible with an IR camera, specifically chosen to obtain a high transmittance to the infrared light. The drum walls are maintained at a desired high temperature using NI-Cr resistive wires and a PID controller. The particle bed temperature evolution is recorded on the IR camera, which is used to determine the average particle bed temperature over time and calculate the wall-to-particle heat transfer coefficient.

630265 Scale-up of Heat Transfer for Rotary Drums with Baffles

Abstract:

Granular materials and powders often require thermal treatment in the food, mineral, pharmaceutical and chemical industries. Rotating drums (kilns) are common devices for the thermal treatment of particulate systems. Developing a better understanding of heat transfer in rotary drums can improve the quality of the product as well as save on energy and material costs. For good product quality, it is often necessary to raise the temperature of the particles uniformly. Although studies have demonstrated the effectiveness and importance of baffles/flights in decreasing the heating time and improving the final product uniformity, many questions about the use of baffles still remain. Understanding the relationship between particle properties and rotary drum operating conditions on the heating time is important for predicting processing time in real-world applications. Simulations using the discrete element method (DEM) were carried out as a means of better understanding the role of baffles in regulating heat transfer. The operating conditions were altered by adjusting the following variables: particle fill level, drum size, baffle size, number of baffles, and the speed of rotation. Furthermore, the effect of material parameters was investigated by varying the size and thermal conductivity of the particles. The results demonstrate how fill level as well as the number and size of baffles play an important role in the heat transfer process. Significant improvements were noted by increasing the number and size of the baffles. The effectiveness of baffles varied as a function of fill level and rotational speed, showing a correlation between these variables. In this work, we also introduced an appropriate characteristic time for baffles that helped us to develop an empirical relationship capable of predicting the average temperature of particle beds.

625132 Continuum Modeling of Corn Stover Feedstock through a Compression Feed Screw

Abstract:

Controllable continuous feeding of biomass through a compression feed screw in an ethanol biorefinery is essential for economic operation. A coupled Eulerian-Lagrangian finite element method model is developed to simulate the feeding of corn stover through a pilot-scale compression feed screw. The bulk corn stover is modeled using an isotropic modified (density-dependent) Drucker Prager Cap (mDPC) constitutive relation. The yield surfaces of the mDPC model capture plastic densification and shear failure of the material. A uniaxial single-ended die compaction test in an instrumented punch-die is used to calibrate compression characteristics, elastic properties, and wall friction properties of biomass. Direct shear tests are performed to partially calibrate shear failure properties of biomass. A sensitivity analysis of compression feed screw performance to mDPC model parameters is presented to assess the possibility of calibrating only a subset of material properties.

624149 Starch Salt Composites; How to Reduce Water Uptake and Weakening

Abstract:

Starch salt composites have been research for a while at our department and we have established composites that match well with the strength of concrete materials. The composite can be seen as an environmental friendly, low carbon footprint building material. This novel building material is easy to make and consist of roughly 90 % NaCl and 10% starch. The compressive strength lies a bit below concrete, but the tensile strength is higher than concrete.

One of the drawbacks it that the material tends to weaken when exposed to water. In this work we have looked into increasing the water resistance of the material by adding additional polymers and by adding chemicals like citric acid that increase the water resistance.

629343 Particle-Level Residence Times in a Twin-Screw Feeder

Abstract:

Residence time distribution (RTD) and flow sheet modeling have proven a valuable tool for material tracking in a pharmaceutical continuous manufacturing process. Feeders are often considered ideal first-in-first-out systems in flow sheets, neglecting the intermixing between batches inside the feeder. In addition, experimental methods to measure the RTD of a discharging feeder are complex and material intensive.

In this work, we present a discrete element model (DEM) of a full discharge of a twin-screw feeder. Feeding is an inherently transient process; there is no steady-state condition. The DEM software XPS (extended particle system, [1]) is capable performing a 16-minute-long discharge process within two months on a single consumer-grade GPU (Nvidia GTX 1080Ti). The initial number of particles was 2.5 million, with the particle number decreasing over time due to discharge. Upon discharge, the starting position, particle radius, and residence time of each particle is logged before the particle is deleted from the system. The log files of individual particle residence times using a free flowing material are available as open dataset [2].

With this dataset it is possible to extract information that is relevant at the process level, for example the feed factors over time. In addition, because the residence time and starting position of each particle are known, it is possible to extract RTDs of arbitrary spatial regions in the feeder. The RTD data per layer enables material tracking inside the feeder for different refill levels [3]. Thus, the particle-level residence time data enable the development of a refill strategy considering both the feed rate and the intermixing of batches inside the feeder.

References

[1] T. Forgber et al., „Extended validation and verification of XPS/AVL-FireTM, a computational CFD-DEM software platform“, *Pow Tech*, **361**, 880–893 (2020). doi:10.1016/j.powtec.2019.11.008.

[2] P. Toson und J. G. Khinast, Dataset: „Particle-Level Residence Time Data in a Twin-Screw Feeder“. *Mendeley Data* (2019). doi:10.17632/D76RZZD8R7.2.

[3] P. Toson und J. G. Khinast, „Particle-level residence time data in a twin-screw feeder“, *Data in Brief*, **27**, 104672 (2019). doi:10.1016/j.dib.2019.104672.

630202 An Experimental Investigation of Ultra-High Temperature and Pressure Fluidized Bed for Thermal Energy Storage and Transfer

Abstract:

Pressurized fluidized beds (PFB) have many applications such as fluidized bed combustion, heat exchange processes, and even to accelerate chemical reactions. A novel energy storage system designed by the National Renewable Energy Laboratory (NREL) utilizes an ultra-high temperature and pressure PFB to transfer heat from particles to air. Experimental data under such conditions are rare in the literature[1], [2], and our work presents data from a laboratory-scale PFB and comparisons to computational models. To validate the computational models, a laboratory-scale PFB was built to operate at 1000°C and 10 bar. The PFB has an inner diameter of 3.125" and the bed height varies from 3" to 9". The particles used were silica sand with a mean diameter of 625 microns. The minimum fluidization velocity, bed pressure drop, and heat transfer coefficient of the PFB were determined from the experimental data. The minimum fluidization velocity and heat transfer coefficient obtained were compared to existing correlations [3]–[6]. Discrete element modeling (DEM) simulations using the Multiphase Flow with Interface Exchanges (MFiX) on the lab-scale PFB were also compared to the experimental results. The models for particle-to-gas heat transfer and fluidization characteristics are validated via comparison to the experiments for a range of temperatures and pressures [1][2].

References

[1] J. Shabanian and J. Chaouki, "Fluidization characteristics of a bubbling gas-solid fluidized bed at high temperature in the presence of interparticle forces," *Chemical Engineering Journal*, vol. 288, pp. 344–358, Mar. 2016, doi: 10.1016/j.cej.2015.12.016.

[2] K. Kitano, S. Chiba, Y. Tazaki, S. Honma, M. Yumiyama, and J. Kawabata, "Minimum fluidization velocity at elevated temperature and pressure," *Kagaku Kogaku Ronbunshu; (Japan)*, vol. 12:3, 1986, Accessed: Apr. 05, 2021. [Online].

[3] ERGUN and S., "Fluid flow through packed columns," *Chem. Eng. Prog.*, vol. 48, pp. 89–94, 1952, Accessed: Apr. 05, 2021. [Online]. Available: https://ci.nii.ac.jp/naid/10003393451.

[4] WEN and C. Y., "Mechanics of fluidization," *Chem. Eng. Prog. Symp. Ser.*, vol. 62, pp. 100–111, 1966, Accessed: Apr. 05, 2021. [Online]. Available: https://ci.nii.ac.jp/naid/20000746320.

[5] Y. Shao, J. Gu, W. Zhong, and A. Yu, "Determination of minimum fluidization velocity in fluidized bed at elevated pressures and temperatures using CFD simulations," *Powder Technology*, vol. 350, pp. 81–90, May 2019, doi: 10.1016/j.powtec.2019.03.039.

[6] D. J. Gunn, "Transfer of heat or mass to particles in fixed and fluidised beds," *International Journal of Heat and Mass Transfer*, vol. 21, no. 4, pp. 467–476, Apr. 1978, doi: 10.1016/0017-9310(78)90080-7.

The use of Two Fluid Modeling to Capture Phenomena Related to Fluid-Solid Transitions in Gas-Solid Fluidized Beds

Qiang Guo[1], Christopher. M. Boyce[1*]

[1]Department of Chemical Engineering, Columbia University, New York, New York 10027, USA

(*) Corresponding author. Email: cmb2302@columbia.edu.

Two fluid modeling (TFM) has been widely used in simulating large-scale gas-solid fluidized beds due to its computational efficiency. However, TFM needs constitutive relations for solids pressure and solids viscosity to model solids stress. Due to the similarity of the physical picture between particle motion in the kinetic regime and thermal motion of molecules in gas, a kinetic theory of granular flows (KTGF) that is analogous to dense-gas kinetic theory is commonly used to correlate the kinetic contribution of the solids stress. In contrast, the theories used to formulate the solids stress in the frictional regime are largely based on the critical state theory of soil mechanics; such efforts, however, are largely empirical as compared to those in the kinetic regime. As a result, TFM with a frictional solids stress model has been shown to struggle to capture some phenomena associated with fluid-solid transitions. Herein, we show that TFM with the Schaeffer model[1] can surprisingly predict granular droplet splitting formed with vibrating the fluidized bed due to local solidification below the droplet[2]. We also show that TFM with conventional frictional solids stress models cannot fully predict structured bubbling pattern in fluidized beds formed with pulsating the gas flow[3] in the bottom inlet or with vibration[4], due to rapid, local transitions between solid-like and fluid-like behavior in the grains. We propose to modify the Srivastava and Sundaresan model[5] by replacing the frictional solids pressure formulation with the one based off a dilation law[6] from prior discrete particle simulations. This new model can successfully predict the structured bubbling pattern formed in fluidized beds with pulsating the gas flow or with vibration.

References:

(1) Syamlal, M.; Rogers, W.; O`Brien, T. J. *MFIX Documentation Theory Guide*; DOE/METC-94/1004; USDOE Morgantown Energy Technology Center, WV (United States), 1993. https://doi.org/10.2172/10145548.

(2) McLaren, C. P.; Kovar, T. M.; Penn, A.; Muller, C. R.; Boyce, C. M. Gravitational Instabilities in Binary Granular Materials. *Proceedings of the National Academy of Sciences of the United States of America* **2019**, *116* (19), 9263–9268. https://doi.org/10.1073/pnas.1820820116.

(3) Wu, K.; Francia, V.; Coppens, M.-O. Dynamic Viscoplastic Granular Flows: A Persistent Challenge in Gas-Solid Fluidization. *Powder Technology* **2020**, *365*, 172–185. https://doi.org/10.1016/j.powtec.2019.04.053.

(4) Guo, Q.; Zhang, Y.; Padash, A.; Xi, K.; Kovar, T. M.; Boyce, C. M. Dynamically Structured Bubbling in Vibrated Gas-Fluidized Granular Materials. *PNAS* **2021**, *118* (35). https://doi.org/10.1073/pnas.2108647118.

(5) Srivastava, A.; Sundaresan, S. Analysis of a Frictional–Kinetic Model for Gas–Particle Flow. *Powder Technology* **2003**, *129* (1–3), 72–85. https://doi.org/10.1016/s0032-5910(02)00132-8.

(6) Pouliquen, O.; Cassar, C.; Jop, P.; Forterre, Y.; Nicolas, M. Flow of Dense Granular Material: Towards Simple Constitutive Laws. *Journal of Statistical Mechanics: Theory and Experiment* **2006**, *2006* (07), P07020–P07020. https://doi.org/10.1088/1742-5468/2006/07/p07020.

632126 The Effects of Adding Soft Particles on the Clogging of Rigid Particles in a 2D Hopper

Abstract:

New directions in cosmetics, pharmaceuticals, and even the food industry increasingly require processing of gel beads and other deformable ingredients. While rigid particle flows through hoppers are fairly well understood, clogging of soft particles and in mixtures of soft and rigid particles is an emerging area of interest. We investigate how the addition of soft hydrogel particles affects the clogging of rigid polypropylene spheres in a quasi two-dimensional hopper both experimentally and through numerical simulations. We also study the characteristics of arch formation at the hopper exit and investigate how the span, aspect ratio, inter-particle angle and number of particles in the arch change with the addition of soft particles. We explore the effect of various parameters by controlling particle softness, particle-particle and particle-wall interactions and hopper geometry, and investigate how they affect clogging probability and arch properties. In our experiments, clogging probability exhibits non-monotonic behavior with respect to the fraction of rigid particles in the mixture, due in part to differences in the contacts of the soft and rigid particles with the hopper wall. In cases of mismatch between the soft and rigid particle sizes, quasi-3d effects may also contribute to the non-monotonicity. The results of our simulations and experiments can help inform improved processing of bidisperse granular flows in many industrial and medical applications.

627297 CFD Simulation of the Entire Circulating Fluidized Bed (CFB) Carbon Capture Loop Using Solid Supported Amine Sorbents

Abstract:

Carbon capture using solid sorbents in considered to be a very promising alternative for aqueous solutions of amine-based absorbents to reduce the cost of regeneration during carbon capture and regeneration processes. The computational fluid dynamics (CFD) approach which is an advanced tool for design and scale up of fluid-particle processes was used in this study to simulate two- dimensional gas and solid flow in a circulating fluidized bed (CFB) that was designed by the National Energy Technology Laboratory (NETL) for carbon capture using solid supported amine sorbents. To gain better understanding of the complex flow patterns of particles and fluid through entire CFB loop, a two-dimensional non-reactive CFD simulations of the entire loop including the carbonator, regenerator, the solid circulation rate and pressure drop at different locations of CFB were performed and the results were compared with the experimental data.

In addition, a two-dimensional reactive simulation of the CFB carbon capture unit was also performed to study the extent of CO_2 adsorption and regeneration using amine-based solid sorbents in the temperature swing carbonation regeneration combined system.

627318 Computational Modeling for FCC RTD Design Optimization

Abstract:

Fluid Catalytic Cracking (FCC) is one of the most important conversion processes in the refining industry. The key to achieve a good performance from the FCC unit highly depends on the internal components, such as riser termination device (RTD), spent catalyst distributor, catalyst cooler, stripper internals etc, In the R2R type of Resid FCC unit, a mushroom type distributor is used to distribute air and spent catalyst into the 2nd regenerator catalyst bed. Uniform distribution is key to achieve a good temperature profile throughout the bed, resulting in better coke burn and improved catalyst activity retention. In PropyleneMax Catalytic Cracking (PMcc) process, a mushroom distributor is used as a riser termination device "RTD". The distributor partially separates the gas and catalyst and distributes the gas into the reactor catalyst bed. Proper distribution of the hydrocarbon vapor into the reactor bed is key to provide proper contact for the vapor to further crack to valuable products. The paper discusses recent developments of Technip Energies mushroom distributor used as a riser termination device in the PMcc reactor and in the RFCC 2nd stage regenerator. The use of CFD modeling to optimize the design will be presented and discussed.

629790 Model-Aided Design and Translation of Glucose-Responsive Insulins: From Simple to Complex Mechanisms

Abstract:

Despite considerable experimental progress, the development of glucose-responsive insulins (GRIs) still largely depends on empirical knowledge and repeated experimentation. Moreover, the only GRI that entered the clinical trials, MK-2640, failed to replicate its outstanding preclinical performance in humans. We developed a program named PAMERAH to assist the rational design and clinical translation of GRIs. PAMERAH constitutes a framework for predicting the therapeutic efficacy of a GRI candidate from its user-specified mechanism of action, kinetics, and dosage, which we show is accurate when checked against data from experiments and literature. Results from simulated glucose clamps also agree quantitatively with recent GRI publications. We demonstrate that the model can be used to explore the vast number of permutations constituting the GRI parameter space, and thereby identify the optimal design ranges that yield desired performance. A design guide aside, PAMERAH more importantly can facilitate GRI's clinical translation by connecting each candidate's efficacies in rats, mice, and humans. The resultant mapping helps find GRIs which appear promising in rodents but underperform in humans. Conversely, it also allows for the discovery of optimal human GRI dynamics not captured by experiments on a rodent population. Our recent efforts focus on applying PAMERAH to more involved GRI mechanisms. We show that MK-2640's complex mechanism of competitive receptor binding and clearance can be modeled equally well as the proof-of-concept kinetics reported earlier. The failed GRI candidate's animal and human design spaces provide insights on why the clinical translation was challenging. On another front, we demonstrate the applicability of PAMERAH to hypothetical GRI designs before any physical prototype is built. By docking PAMERAH with circuit simulations, we demonstrate the feasibility of a microrobotic GRI where the insulin release is intelligently controlled by a printed cell-sized circuit.

632092 Programmable Biohybrid Nanocarriers for the Sustained Release of Cancer Therapeutics

Abstract:

Cancer is the second leading cause of death worldwide. It is estimated that 1.9 million new cases and 608,570 deaths due to cancer will emerge in 2021 alone. Due to the excessive costs of cancer diagnosis and treatment the projected costs will reach $246 billion by 2030 in the US. As a result, the quest for a cost-effective cancer therapy with broad applications across cancer and patient types has garnered significant interest. Two of the most impactful innovations in the fight against cancer in the past few decades have been immune therapy and polymeric nanocarriers. Using these strategies, effective treatments have been developed that can trigger the patient's immune system to attack cancer cells or accumulate therapeutics in cancerous tissue through the enhanced permeability and retention effect (EPR). However, due to drug resistance mechanisms unique to cancers, a number of recalcitrant cancers still have poor prognosis and metastasis lowers survivability for all cancer types. Therefore, an elegant nanocarrier - with high drug payload, delayed or triggered release, and tailorable targeting components – has the chance to revolutionize modern cancer treatment. The availability of commercially synthesized oligonucleotides has flooded the field with novel nucleic acid nanostructures, therapeutics, and biosensing elements (including the excellent molecular targeting capabilities of aptamers). The invention of spherical nucleic acid particles created a new class of nanocarrier whose potential is still being realized. Evidence has shown that specifically oriented oligonucleotides on spherical nanoparticle supports are highly stable, resistant to opsonization, produce a lesser immune response than free nucleic acids, and are internalized into cells without an additional transfection reagent. Additionally, work from our lab has pioneered sustained release from nucleic acid polymers through affinity modulation, only possible for small molecules through careful selection of the nucleic acid sequence (i.e., aptamer recognition). Thus, by tailoring the architecture of our novel biohybrid nanocarrier, we are able to program the particle according to the design principles for creating effective therapeutic nanocarriers: bioavailability, payload retention, and targeted delivery. We coat a 15 nm gold nanoparticle (AuNP) with short, duplexed oligonucleotides (dsAGC) chosen for their high affinity toward the chemotherapeutic daunomycin. A single stranded oligonucleotide (ssANC) is then conjugated alongside the dense duplex layer by way of a charge-neutral PEG spacer and can be used to hybridize additional oligonucleotides. The completed platforms are less than 100 nm in diameter and are resistant to serum protein adhesion compared to similar particles made with only single-stranded DNA. Each dsAGC oligonucleotide is able to bind 3-4 molecules of daunomycin and they are conjugated at a density of greater than 100 oligonucleotide per AuNP. The high affinity of daunomycin to the AGC motif results in a sustained release of daunomycin over more than 48 hours. Additionally, we show that dexamethasone can adhere directly to the gold surface and that the release rate can be controlled by increasing the density of DNA on the AuNP. In conclusion, we have developed a novel, biohybrid nanoparticle that can load and release at least two distinct chemotherapeutics. The high DNA density and structure of the resultant particles imbue a resistance to protein adhesion, likely due to steric hindrances. The small size of the particles will allow them to exploit the EPR effect of solid tumors. Additionally, the ssANC oligonucleotide can hybridize with complementary sequences making them programmable with targeting aptamers, peptides, or monoclonal antibodies, providing active targeting capabilities. As personalized medicine techniques provide improvements to active cancer targeting, we expect that our design will provide an essential platform for the delivery of chemotherapeutics to a broad range of cancer types.

632604 Self-Organization of Iron Sulfide Nanoparticles into Multi-Compartment Supraparticles

Abstract:

Self-organization of Iron Sulfide Nanoparticles into Multi-compartment Supraparticles

Emine S. Turali-Emre[1,5], Ahmet E. Emre[1,5], Luis Rivera-Rivera[2,5], Usha Kadiyala[4,6], Thi Vo[2,6], Scott VanEpps[4,5,6], Sharon C. Glotzer[2,3,5], and Nicholas A. Kotov[1,2,3,5,6*]

[1]Biomedical Engineering Department, University of Michigan, Ann Arbor, USA;

[2]Chemical Engineering Department, University of Michigan, Ann Arbor, USA;

[3]Materials Science and Engineering Department, University of Michigan Ann Arbor, USA;

[4]Department of Emergency Medicine, University of Michigan, Ann Arbor, Michigan 48109, United States

[5]Macromolecular Science and Engineering Department, University of Michigan, Ann Arbor, USA;

[6]Biointerfaces Institute University of Michigan; University of Michigan; Ann Arbor, USA

Gene and gene editing therapies have been widely investigated for treatment of inherited or acquired genetic diseases. Efficient delivery of therapeutic agents has become a significant barrier in clinical applications due to the toxicity and instability of the vectors in the complex intracellular environment. Among non-viral vectors, inorganic nanoparticles (NPs) have become a popular strategy for nucleic acid delivery. Even though, NPs can condense the nucleic acids, the nanoshell geometry of viruses is advantageous for the gene/gene editing therapies cargo protection. Therefore, we synthetized iron-based inorganic nanoparticles which self-assemble into supraparticles with nanoshell geometry. We synthetized L-cysteine stabilized iron sulfide NPs in aqueous media to eliminate toxic reagents during synthesis. Careful selection of the reaction conditions such as precursor concentrations, pH level and reaction time allowed us to find a condition where iron sulfide NPs self-assembly into virus-like supraparticles. Transmission electron microscopy (TEM), TEM tomography and dynamic light scattering (DLS) were used to characterize these supraparticles for size, shape, and charge. Our results indicate that virus-like supraparticles contain continuous compartments, are positively charged (25±7.2 mV) and 74±21 nm in diameter. We loaded deoxyribonucleic acid (DNA) in the compartments during the formation of supraparticles. We tested these complexes in circular dichroism, UV-Vis spectroscopy, electrophoretic mobility shift and protection assays to confirm the encapsulation of DNA in the compartments. Since iron sulfide is a natural material, it presumably has low cytotoxicity and high biocompatibility. Supraparticles can condense DNA, protect it against degradation, penetrate through cellular membranes and facilitate endolysosomal escape in gene therapy. Therefore, development of these virus-like particles can be used as an effective cargo delivery tool for gene and gene editing therapies such as CRISPR.

Understanding die compaction of hollow spheres using the multi-particle finite element method (MPFEM)

Abstract ID# 623006

Ahmet Demirtas[a*] and Gerard R Klinzing[a]
[a] *Merck & Co. Inc., Kenilworth, New Jersey 07033*
*Corresponding author e-mail address: ahmet.demirtas@merck.com

Powder compaction is a complex manufacturing process. Although the procedure follows four presumably simple steps: (*i*) die filling, (*ii*) compaction, (*iii*) unloading, and (*iv*) ejection, it is a very challenging process to understand due to its complex nonlinear micromechanical and statistical nature. It is especially challenging for empirical studies to investigate particle-level interactions. Thus, computational analyses are required for particle-level understanding. A wide range of computational methods has been developed, grouped based on the macromechanical approach (continuum approach) and based on the micromechanical approach (discrete element methods). Although the continuum approach can provide information on the evolution of tablet density and stress upon compaction, unloading, and ejection, it ignores the problem's discrete nature. The discrete element method (DEM) and the multi-particle finite element method (MPFEM) were developed to understand the discrete behavior and the global behavior of the powder compaction.

DEM technique successfully represents the particle level interactions; however, it cannot compensate for the deformed particle surfaces during compaction, which plays an essential role in determining the contact pressure between contact pairs (particle-particle and particle-tooling). Although new approaches were developed in recent studies to address the issues, these approaches require particular calibration methods, which increase the computational cost. MPFEM addresses the challenges associated with DEM and other macromechanical techniques by incorporating the characteristics of both methods. Although MPFEM helps to identify the particle level interactions, the computational load is heavy. Due to the computational limitations, a limited number of studies in the literature have analyzed the powder compaction using a 3D MPFEM. Historically, these studies focus on solid particles.

In this study, hollow spheres' compaction behavior, common to pharmaceutical spray drying, was investigated both computationally and experimentally. In the computational analysis, two different particle sizes with three different shell-thickness levels were examined using the 3D MPFEM via a commercially available FEM software Abaqus (Version 2019, Simulia, Providence, RI). Particles were grouped into two based on their size: (*i*) Large particles with a unit radius (r_L) and (*ii*) small particles with a 25% smaller radius (r_s). The initial particle coordinates were generated randomly using a simple DEM particle generator simulation, which was only allowed to generate one type of particle at a given time. DEM simulations run until the first 2000 particles were settled in the die, i.e., the average displacement in all particles reached a plateau. Keeping the particle numbers under 2000 allowed to keep wall clock time for the simulations to <60hrs. Additionally, the total mass of models was kept constant for consistency. The models were divided into groups based on the outer particle radius and by the particle diameter (d): shell-thickness (w) ratio. In this study, the thick shell, medium shell, and thin shell particles were represented with a d/w of 2.5, 5, and 10, respectively (**Table 1**). The particles were designed as elastic-perfectly plastic with E/Σ_y of 100 and Poisson's ratio (v) of 0.3, where E and Σ_y are Young's modulus and the uniaxial yield stress, respectively. Since material property selection directly influences the stable time increment the accuracy of the results, an optimized selection must be made on material properties. All simulations run up to approximately 0.9 *RD*. The punch velocity was set to complete the compaction stage in 5 seconds.

The spray-dried intermediate (SDI) used for the experimental portion of this study was produced on a Mobile Minor™ spray dryer (GEA Niro A/S, Denmark). The polymer used was hydroxypropyl methylcellulose acetate succinate (HPMCAS) M-grade from Ashland (Wilmington, DE). Compaction experiments were performed using a Huxley-Bertram (Cambridge, U.K.) hydraulic compaction simulator with round flat-faced punches with 2mm punches. Each tablet was compressed after weighing approximately 4-4.5 mg of powder. The powder was manually fed into the die. The die fill height was set to 16 mm to allow for the greatest volume for powder fill. The compaction process was completed in 10 seconds, including 5 seconds for compression, 5 seconds for unloading, and 2 seconds for ejection starting after unloading was complete. The compaction requirements were satisfied with a single action movement until the total punch separation between the upper and lower punch was 1 mm. Following the tablet formation process, weight, thickness, and diameter measurements were executed immediately. Each tablet was weighted using a Mettler Toledo balance with a 0.1 mg resolution, and the thickness and diameter were measured with a Mitituyo thickness gauge with a resolution of 1 μm modified with a flat contact surface for the measurement probe. Uncompressed SDI particles and tablets were scanned with a high-resolution XRadia Versa 500 (White Plains, NY, USA) x-ray tomography. Surface characterization of SDI particles and tablets was performed on an FEI Quanta FEG 250 (FEI Corporate, Hillsboro, OR, USA) environmental scanning electron microscope.

The results showed that particle diameter/shell-thickness (*d/w*) plays an essential role in powder compaction behavior. The initial relative density, the relative density after die filling but prior to compression, and the fill depth varied with respect to the particle diameter and *d/w* ratio. During the initial stage of compaction, the punches remained stress-free, or the stress was negligible, despite the non-zero velocity of the upper punch at this early stage of compression. Once the stress levels are not negligible, a steady stress increase was observed in all models (Figure 1a). As compression continues, the rate at which stress increases is highly dependent on the particle's *d/w* ratio. The *d/w* ratio 2.5 models show the steepest increase in compaction stress. The *d/w* 5 particles show a softer response up to the point at which the relative density equals that of the *d/w* 2.5 particles at a Σ_y=0.58 and *RD*=0.78 (Figure 1a). At this point, the two particles begin to behave similarly up through the end of compression, where the d/w 5 particles show slightly lower peak stress. The response of the thin-shelled particles, *d/w* 10, shows drastically different results across the entire compaction range. Upon final compression, the *d/w* 10 particles do not reach the same stress as achieved by the *d/w* of 5 and 2.5 despite reaching the same final *RD*. However, based on the stress trajectory of the *d/w* 10 models, it is expected to reach stresses equivalent to the other models but not until an *RD* higher than 0.9 is achieved. It was also found that shell-thickness is the dominant factor in particle yielding (Figure 1b). The yielded volume was calculated by summing the total volume of the elements within all particles that exhibited non-zero equivalent plastic strain. Since the selected material property is elastic-perfectly plastic, plastic strains on deformed elements do not increase local particle stresses. The total yield volume (*YV*) was normalized by the total particle volume (*TV*) showing yield density in a model, e.g. if the ratio (*YV/TV*) is equal to 1, all the particles in the model passed the yield point. Although the particle size is shown to alter the global yielding behavior, the influence is not as dominant as the shell-thickness. The results show that thin-shell models tend to reach total yielding at much lower macroscopic stress than the thick- and medium-shell models. Also, small particles require slightly higher forces to have the same amount of yield before reaching complete yielding because of the larger contributions of inter-particle friction per unit volume (Figure 1b).

Experimental compaction studies were performed using the two different SDI particles produced. The particles generated with the higher outlet temperature were larger in diameter and thinner in wall thickness than the particles generated with the lower outlet temperature. The larger particles with thinner shells (*SDI*-80 *°C*) initially build stress upon consolidation greater than smaller thicker shelled particles (*SDI*-45 *°C*) (Figure 1c). This is a direct result of the difference in particle packing between the two SDI particles. However, the smaller, thicker shelled particles result in a stress value greater than that of the larger thinner particles for equivalent RD upon further compression. This trend is consistent with that of the MPFEM model (Figure 1c-d). As the compression event continues, the thinner shelled particles start to show a softened response, as was observed experimentally. This is likely due to the buckling phenomenon of the thinner shell larger particles. Upon compression to an *RD* approaching 0.9, the MPFEM simulation stresses continue to show agreement with the experimental results. As was captured in MPFEM, the required stress to reach the final compression level (*RD*= 0.9) was higher for the thick-shell particles than the thin-shell particles. Due to the lack of cohesive elements used in this work's modeling portion, the MPFEM analysis was only performed through the compaction and unloading event. Even though the final MPFEM tablet upon ejection could not be compared, there are interesting similarities between the unloaded MPFEM models and the physical tablets (Figure 1e-f).

In summary, this study provides new information on how powder compaction behavior was influenced by particle size and particle shell-thickness. The approach used in this study helps to understand the complexity involved in the powder compaction process and may be used to identify the additional mechanisms during the compaction process.

Table 1: The detailed description of finite element models used in the study.

Model Name	**Number of Particles**	**Description**
r1-2.5	411	Each particle has a unit length radius, a diameter shell-thickness ratio of 2.5. The initial total mass is constant.
r1-5	520	Each particle has a unit length radius and a diameter shell-thickness ratio of 5. The initial total mass is constant.
r1-10	835	Each particle has a unit length radius and a diameter shell-thickness ratio of 10. The initial total mass is constant.
r0.75-2.5	973	Each particle has a 75% unit length radius and a diameter shell-thickness ratio of 2.5. The initial total mass is constant.
r0.75-5	1232	Each particle has a 75% unit length radius and a diameter shell-thickness ratio of 5. The initial total mass is constant.
r0.75-10	1979	Each particle has a 75% unit length radius and a diameter shell-thickness ratio of 10. The initial total mass is constant.

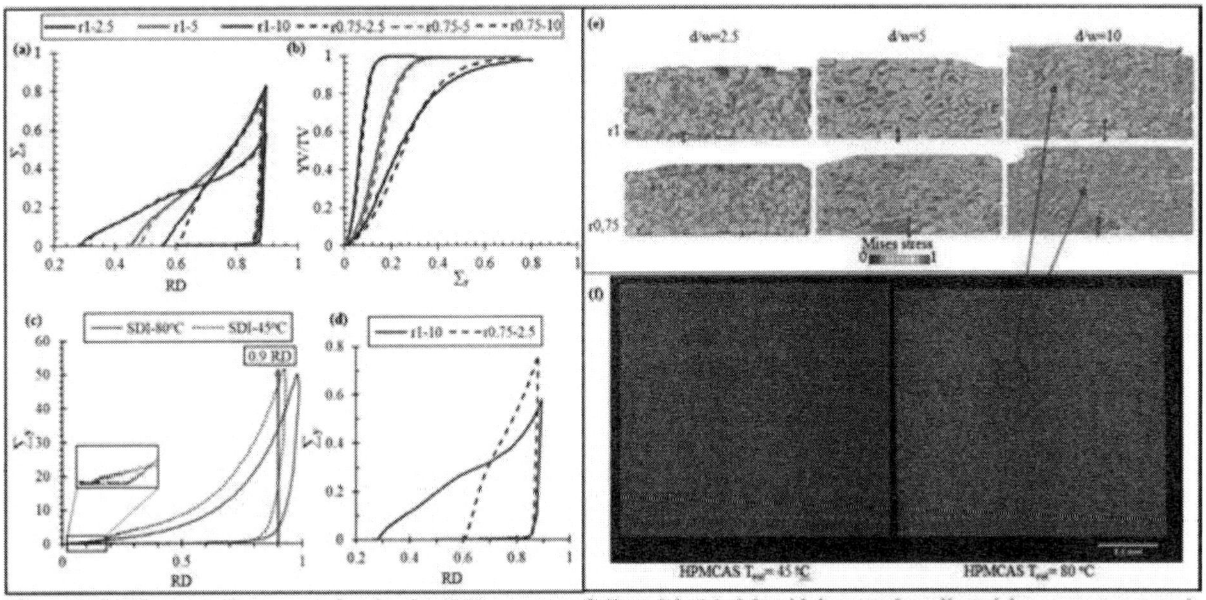

Figure 1: (a) The stress (Σ_y) vs. relative density (*RD*) response of all models. Models with large and small particles were represented with solid and dashed lines, respectively. The *d/w* ratio of 2.5, 5, and 10 was illustrated with black, red, and blue colors, respectively. (b) The ratio of yielded element volume (YV) to the total elemental volume (TV) shown as a function of stress. (c) Experimental and (d) MPFEM stress (Σ_y) vs. relative density (*RD*) response. (e) Simulation and (f) experiment release stage comparison between. (e) From left to right, each column is for a d/w ratio of 2.5, 5, and 10. The top row, marked with r1 on the left side, is the cross-sections of models with large particles, and the bottom, marked with r0.75, is the cross-sections of models with small particles. The red vertical lines with diamonds on both ends show the recoil distance on each model. Note that the contour is the Mises stress distribution. (f) The concave and convex curves are shown with red-colored dash lines. The marked blue region shows the buckled particles in experiment and simulation.

630532 Access: Autonomous Characterisation and Calibration Using Evolutionary Simulation Software

Abstract:

The discrete element method (DEM) is a powerful simulation technique that is capable of numerically modelling the behaviour of complex granular media, being used to better understand and optimise the internal dynamics of a large number of systems in both academic fields and industrial sectors, from fundamental research into contact mechanics to improving plant-scale reactors [1]. DEM can offer exceptional accuracy through its lack of approximations over meshes and, if correctly calibrated, simulations can provide results with quantitative precision. It is this "if", however, that also represents DEM's biggest drawback: without choosing appropriate contact models and carefully calibrating multiple DEM parameters, the simulation outputs simply cannot be trusted. This calibration is a time-consuming process, typically involving the measurement of diverse particle properties including size, density, restitution and friction coefficients and, for purely "virtual" parameters such as the *cohesive energy density*, a great deal of experimentation [2].

To automate DEM calibration against experimental measurements, we have developed ACCESS – Autonomous Characterisation and Calibration using Evolutionary Simulation Software. ACCESS enables a researcher to calibrate virtually *any* DEM parameters against a user-defined cost function, quantifying and subsequently minimising the disparity between the simulated system and the experimental reality using state-of-the-art evolutionary strategies – in essence, autonomously 'learning' the physical properties of the particles within the system, without the need for human input. This cost function is completely general, allowing ACCESS to calibrate DEM against measurements as simple as photographed occupancy plots, or complex system properties captured through e.g. Lagrangian particle tracking. In our tests, ACCESS successfully calibrated up to 12 unknown particle parameters in a matter of hours against the free surface shape of an industry-standard GranuTools GranuDrum™.

ACCESS also represents a solution to the open problem of efficiently simulating large systems – the autonomous determination of particle properties is agnostic to the size of the particles simulated relative to the true size of particles used in experiment. Thus, it can be used to directly determine optimum properties for coarse-grained particles, which by definition cannot be calculated from direct characterisation measurements. A calibrated coarse-grained model would allow a typical simulation to be performed with a reduction in computation time between 160x and 1800x (simulation software dependent) compared to a full, non-coarse-grained simulation.

The algorithm itself is completely DEM engine-agnostic; it was implemented in an open-source Python library, providing an interface that is easy to use, but powerful enough to automatically parallelise arbitrary user scripts through code inspection and metaprogramming. It was used successfully from laptop-scale shared-memory machines to multi-node supercomputing clusters. Though ACCESS was built as a tool for the computational granular mechanics community, it is a fully fledged massively-parallel evolutionary optimisation framework that can work - without any modifications! - with arbitrary simulation techniques that require calibration against experimental data.

REFERENCES

[1] Rosato AD, Windows-Yule C. Segregation in Vibrated Granular Systems. Academic Press; 2020 Jun 5.

[2] Luding S. Introduction to discrete element methods: basic of contact force models and how to perform the micro-macro transition to continuum theory. European journal of environmental and civil engineering. 2008 Aug 1;12(7-8):785-826.

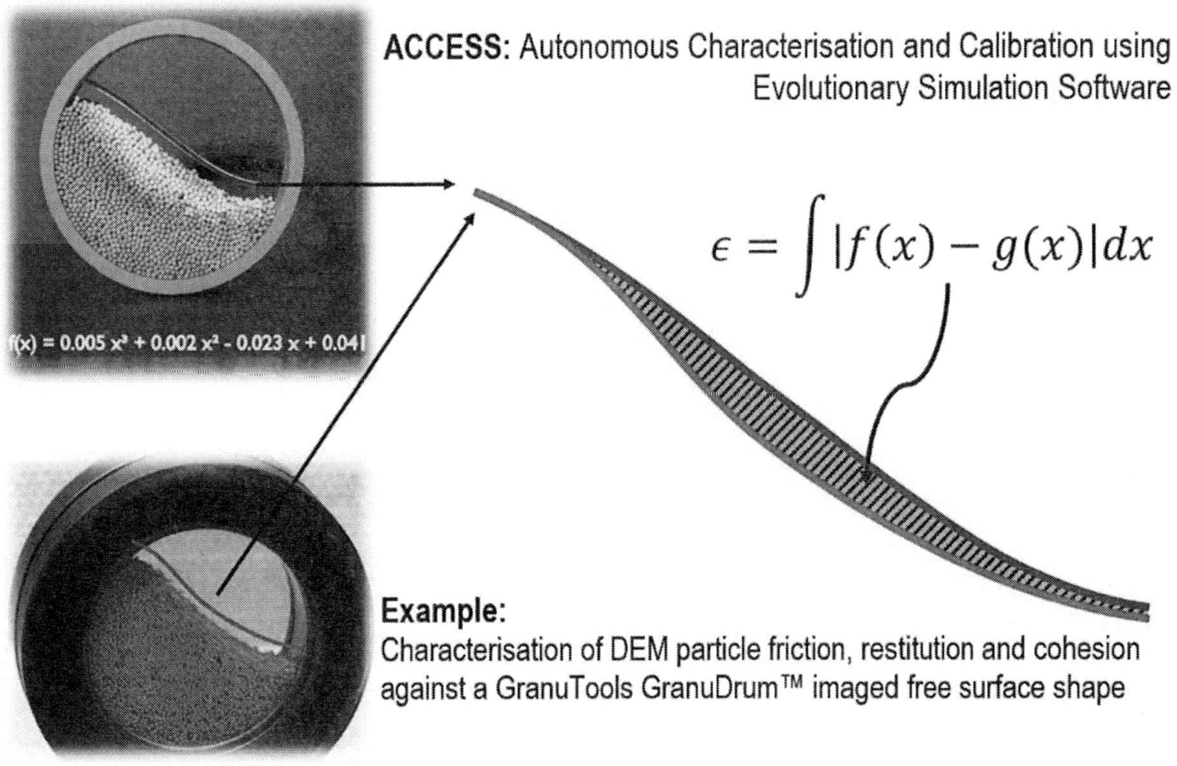
ACCESS: Autonomous Characterisation and Calibration using Evolutionary Simulation Software
f(x) = 0.005 x³ + 0.002 x² - 0.023 x + 0.041
$\epsilon = \int |f(x) - g(x)| dx$
Example:
Characterisation of DEM particle friction, restitution and cohesion against a GranuTools GranuDrum™ imaged free surface shape

624060 Machine Learning-Based View Factor Modelling in Polydisperse Particle Beds Including Walls

Abstract:

Introduction

Radiative heat transfer is the most dominating heat transfer mechanism above 700 [°C]. Such high temperature appear in a wide range of industrial processes: pebble bed reactors, laser sintering and high-temperature particle oxidation, or reduction processes [1]. If no participating fluid is involved, or the contribution of this fluid is neglectable, radiative heat transfer in these processes is typically modelled via surface-to-surface radiation. A prominent approach for surface-to-surface radiation modelling is to calculate the heat flux based on view factors. A view factor (in literature also named 'configuration factor') is the ratio of the radiation leaving an emitting surface and the radiation that is striking an absorbing surface. Industrial scale processes feature wide particle size distributions. Due to the large number of particles it is computationally infeasible to numerically integrate view factors based on first principles, or use Monte Carlo-based algorithms [2] for their calculation. Also, literature methods that aim on decreasing the computational effort [3] are still too demanding to be generally adopted for polydisperse systems. On top of that - except for a single study [4] - wall effects are typically neglected.

Methods

The usage of Machine Learning-methods especially Deep Neural Networks (DNNs) has significantly emerged in simulations science fields like Computational Fluid Dynamics (CFD) and DEM. Also modelling radiative heat transfer and view factors comes into focus [5,6]. In our previous work [7] we tackled the above described fundamental issues of wall effects and exuberant computational demand also by using DNNs. For monodisperse systems the created DNNs were able to accurately predict view factors among particles, as well as between particles and walls at comparably low computational cost.

To solve the remaining issue of polydispersity we create a dataset of view factors that considers a wide range of particle radii ratios, different total particle volume fractions, as well as different particle-species volume fractions. In a first step we directly apply our DNNs modelled for monodisperse interactions to polydisperse interactions and investigate the effects on the view factors distribution and input data. Subsequently, we investigate the effect of the following additional input features of DNNs for polydisperse packings: (i) variants of the size ratio between the interacting objects, (ii) information about the particle volume fraction, and (iii) statistics about the (size) species distribution. Feature selection is typically based on correlation coefficients like the Pearson and Spearman coefficient. For the usage of pretrained models in, e.g. DEM simulations, additional aspects like the computational calculation cost and independency on the system size are important. Therefore, the calculated correlation coefficients are set into relation to the other mentioned aspects and the most promising input markers are selected.

Beside DNNs other Machine Learning methods like (i) decision tree-based random forest regression [8], and (ii) gradient boosting (XGBoost) machines for regression [9] look promising for view factor modelling. Therefore, our DNN-based models are compared to these other regressors. All three regression methods rely on the same selected input features and the same cost function. To achieve the best possible generalization for the DNNs, Random Forest regressor, and XGBoost-regressor, all model parameters are optimized by a randomized search approach [10]. The overall quality of prediction is measured by the mean squared error (*MSE*), which also represents the cost function of the regression. Since the is significantly more influenced by large view factors, and does not reflect the overall quality of the view factor prediction, the coefficient of determination (R^2) between the predictions and the targets is used as additional metric. Another quantitative measure is the coefficient of determination ($R^2_{spread}(x_u,x_o)$) of particle distances within an upper boundary x_o (set to half the domain size plus the mean particle diameter), and a lower boundary x_u (set to half the domain size minus the mean particle diameter).

Results and Conclusion

As expected the direct application of DNNs trained for monodisperse interactions does not yield satisfying results, since monodisperse and polydisperse view factor distributions show significant differences. However, it is shown that the addition of particle size information to monodisperse DNNs significantly improves the performance in polydisperse beds. These adjusted DNNs allow the same prediction quality for polydisperse systems as in monodisperse systems. The latter can be stated for particle-particle as well as particle-wall view factors, although particle-wall view factors show different view factor distribution than particle-particle interactions.

This different data structure also explains the twofold results when comparing between the three ML-methods. For particle-particle view factors the DNN yield the best results with an *MSE* of 5.299e-7, R^2 of 0.9806 and R^2_{spread} of 0.8056. The *MSE* of the Random Forest regressor and the XGBoost-regressor are 5.254e-7 and 4.160e-07, the R^2-values are 0.9808 and 0.9848 and R^2_{spread} is 0.1055 and 0.2004, respectively. Thus, especially the DNN-based prediction of the view factor deep in the particle bed, quantified by , is significantly better than the other two regressors. For particle-wall view factors the Random Forest Regressor is clearly superior. The *MSE* is 1.308e-11, R^2 is 0.999 and R^2_{spread} is 0.999. The *MSE* of DNN-based regressor and XGBoost-regressor are 1.106e-7 and 3.139e-7, the R^2-values are 0.9772 and 0.9353 and R^2_{spread} is 0.6027 and 0.2373.

In terms of overall performance, the DNNs show a higher general applicability, while for specialized tasks other Machine Learning methods show a better performance, e.g. Random Forest regression for particle-wall view factors. The DNNs also benefit from being easier to be included into simulation frameworks like CFD or DEM. Thus, a combination of two DNN-based models, one for particle-particle interactions and one for particle-wall interactions, is recommended for general heat radiation modelling in DEM simulations.

References

[1] M.F. Modest, Radiative Heat Transfer - Second Edition, Academic Press, 2003. https://doi.org/10.1017/CBO9781107415324.004.

[2] T. Walker, S.C. Xue, G.W. Barton, Numerical determination of radiative view factors using ray tracing, J. Heat Transfer. 132 (2010) 1–6. https://doi.org/10.1115/1.4000974.

[3] T. Forgber, S. Radl, A novel approach to calculate radiative thermal exchange in coupled particle simulations, Powder Technol. 323 (2018) 24–44. https://doi.org/10.1016/j.powtec.2017.09.014.

[4] E.F. Johnson, İ. Tarı, D. Baker, Radiative heat transfer in the discrete element method by distance based approximations of radiation distribution factors, Powder Technol. (2020). https://doi.org/10.1016/j.powtec.2020.11.050.

[5] H.H. Kang, M. Kaya, S. Hajimirza, A data driven artificial neural network model for predicting radiative properties of metallic packed beds, J. Quant. Spectrosc. Radiat. Transf. 226 (2019) 66–72. https://doi.org/10.1016/j.jqsrt.2019.01.013.

[6] H. Wu, N. Gui, X. Yang, J. Tu, S. Jiang, A matrix model of particle-scale radiative heat transfer in structured and randomly packed pebble bed, Int. J. Therm. Sci. 153 (2020) 106334. https://doi.org/10.1016/j.ijthermalsci.2020.106334.

[7] J. Tausendschön, S. Radl, Deep Neural Network-based heat radiation modelling between particles and between walls and particles, *Submitted Paper*, (2021).

[8] L. Breiman, Random Forests, Mach. Learn. 45 (2001) 5–32.

[9] T. Chen, C. Guestrin, XGBoost: A scalable tree boosting system, Proc. ACM SIGKDD Int. Conf. Knowl. Discov. Data Min. 13-17-August-2016 (2016) 785–794. https://doi.org/10.1145/2939672.2939785.

[10] J. Bergstra, Y. Bengio, Random search for hyper-parameter optimization, J. Mach. Learn. Res. 13 (2012) 281–305.

624401 Transition between Shear-Induced Segregation and Free-Sifting of Fines

Abstract:

Granular mixtures of particles with various properties (e.g., size, density, shape, and roughness) tend to segregate when subjected to mechanical excitation in a wide range of industrial processes and nature. For a mixture of size-bidisperse spherical particles, the large to small particle diameter ratio, R, determines whether shear-induced segregation (small R) or free-sifting (large R) occurs, but the transition between these two extremes remains largely unexplored. Using discrete element method simulations, we study percolation of small particles (fines) in static random packings of large particles and in free-surface flows with R ranging from 3 to 7.5. For the static case, there are three regimes. (i) For $R>6.464$ (the geometric trapping threshold, which characterizes the smallest opening formed by identical large particles), all fines fall through the voids between large particles due to gravity alone, and the average percolation velocity is constant and nearly independent of R. However, the percolation of fines is normally diffusive only for $R>6.6$ but anomalous for $6.464<R<6.6$. (ii) A transition occurs for $4<R<6.464$, where only a fraction of the fines percolates in a static bed of large particles, taking longer to do so and percolating an increasingly shorter distance as R is reduced. (iii) For $R<4$, particles do not appreciably penetrate the static bed. For the free-surface flow case (periodic chute flow), fines for $R>5$ percolate quickly (compared to the static bed) at a velocity that is only weakly dependent on R and the local shear rate, much like free-sifting in a static bed. For $R<5$ the small particle percolation velocity decreases with decreasing R, approaching zero in the monodisperse limit.

623067 Designing Non-Segregating Granular Mixtures for Free Surface Flows

Abstract:

Granular materials tend to segregate due to differences in particle size and density. However, for bidisperse mixtures of large/heavy and small/light spherical particles, the effects of these differences can be balanced for specific combinations of particle size ratio, density ratio, and mixture concentration so that the particles remain mixed. Using discrete element method (DEM) simulations of a quasi-two-dimensional rotating tumbler we show that segregation is minimized for initially well-mixed particle mixtures near the equilibrium conditions. More significantly, we show that initially segregated particles mix and remain mixed. Overall, our results demonstrate that in free-surface flows, non-segregating granular mixtures can be realized through principled design. Funded by NSF Grant CBET-1929265.

633993 Deep Composites: Bioinspired AI towards Modeling, Design and Manufacturing

Abstract:

Nature produces a variety of materials with many functions, often out of simple and abundant materials, and at low energy. Such systems - examples of which include silk, bone, nacre or diatoms - provide broad inspiration for engineering. Here we explore the translation of biological composites to engineering composites, using a variety of tools including molecular modeling, AI and machine learning, and experimental synthesis using 3D printing, and characterization. We review a series of studies focused on the mechanical behavior of materials, especially fracture, and how these phenomena can be modeled using a combination of molecular dynamics and machine learning. We present examples that involve deep convolutional neural networks, transformer neural networks, and game theoretical approaches towards design. One case study will cover a recent example that realizes a text-to-material design approach, developing new architected multimaterial composite designs based on human readable description and subsequent 3D printing – from word to matter. We conclude the talk with case studies of material optimization using genetic algorithms, applied to 3D printed composites, protein design, and a translation of protein folding to music and back to assess universal patterns.

Evaluation Of Process Variable Influence on Mechanical Properties of Short Fiber-Reinforced 3D-Printed Parts

Authors:

Martin Etemadi; Virginia Tech, Blacksburg, VA, USA
Arit Das; Virginia Tech, Blacksburg, VA, USA
Bradley A. Davis; Hume Center for National Security and Technology, Blacksburg, VA, USA
Steven H. McKnight; Virginia Tech Alexandria Campus (National Capital Region), Alexandria, VA, USA
Scott W. Case; Virginia Tech, Blacksburg, VA, USA
Christopher B. Williams; Virginia Tech, Blacksburg, VA, USA
Michael J. Bortner; Virginia Tech, Blacksburg, VA, USA

Abstract

Fused filament fabrication (FFF) has seen broad industrial adoption as it is capable of manufacturing large complex parts from robust thermoplastics in a cost-effective manner. However, the mechanical performance of the printed parts is limited due to poor interlayer bonding and the presence of voids. In order to overcome these drawbacks, the addition of short or continuous fibers into the polymer matrix has been investigated, as the fibers can act as a mechanical reinforcement while also mitigating residual stress resulting from the material's rapid solidification following extrusion. Therefore, understanding the implications of process parameters and fiber reinforcements on printed part properties through detailed crystallization analysis and rheological characterizations is of paramount importance.

The goal of this study is to understand the process–structure–property relationships of short carbon fiber reinforced polyamide 6 (CF-PA6) by comparing the melt rheology and crystallinity of CF-PA6 versus a neat PA6 polymer. Differences in the melting and crystallization behavior resulting from the reinforcing fibers revealed an increased time window for crystallization in the fiber-reinforced matrix. Rheological characterizations at the recommended printing temperatures demonstrate the shear-thinning behavior of the samples at shear rates relevant to FFF. From a statistical design of experiments analysis, the layer thickness was found to be the most significant parameter affecting the tensile properties of a printed part at a constant printing temperature and printing speed. The tensile fracture surfaces of the printed specimens using scanning electron microscopy were analyzed to provide insights into the failure mechanisms as a function of AM processing variables.

KEYWORDS:

3D printing, additive manufacturing, composites, polyamides, rheology

622245 Direct Ink Write of High Solid Suspensions: Considerations in Particle Type and Binder Properties

Abstract:

Construction, energetics, and regenerative medicine are several fields which rely on processing of suspensions with high content of solid particles (> 60 vol%). New processing methods for these areas, such as direct ink write (DIW) 3D printing, have enabled greater part complexity, reduced material waste, and improved safety during manufacturing. However, maximizing solid content presents processing difficulties due to the limited amount of binder in these suspensions, namely high viscosities, poor print resolution, and heterogeneous particle distributions. Despite these complications, for systems with particle diameters on the order of 10^2 µm, binder effects become more pronounced due to larger interstitial spaces between particles, but these effects remain poorly understood. Providing insights into the complex binder-particle and particle-particle interactions that occur at this scale is crucial for the development of successful DIW formulations. Here, we investigate the dual effects of particle type as well as binder factors in two different solidification mechanisms: solvent evaporation and UV cure. Using rheological tests, we characterize the viscoelastic response of these inks to applied stress mimicking the extrusion process. We use oscillatory time and 3 Interval Thixotropy experiments to analyze solidification post-printing as well as structure breakdown during shear followed by regeneration. We also visualize the microstructure prior to printing and after curing through scanning electron microscopy. For particles, we test spherical SiO_2, non-uniform SiO_2, and non-uniform basaltic soil particles. Binder factors of interest are surface tension of solvents and molecular weight and glass transition temperature of polymers. Fundamental understanding of the interplay between these components during 3D printing will enable a high level of customization of DIW inks for successful processing, enabling precise geometries tailored for novel buildings, safer energetics, or personalized medical devices.

632114 Dimensionalization of Two-Phase Newtonian/Non-Newtonian Flow Problems.

Abstract:

Non-Newtonian fluids are present in a large variety of everyday life applications including paints, foods, adhesives, inks, cements, slurries, polymers and gels. Computational fluid dynamics (CFD) simulations have been widely used to understand, optimize and design products and processes that involve such non-Newtonian fluid flows. Further complicating such challenges are cases wherein flows involve the interactions of multiple phases. Systematically, understanding complex flows is difficult since they generally involve multiple forces including inertial, viscous, gravitational, capillary forces. The present study suggests dimensionless forms for the momentum and continuity equations along with two-phase level-set and phase field equations for non-dimensionalizing two-phase, non-Newtonian fluid flow problems. The Reynolds number, Galilei's number and Capillary number arise naturally as scaling factors. Fluid dynamic (CFD) simulations using COMSOL Multiphysics® were used to illustrate comparative numerical analysis between dimensional and dimensionless equation forms for two-phase, non-Newtonian pipe flow and free-surface flow (slump-flow) examples. The effect of non-dimensionlization on computational time was demonstrated for different scale of free-surface flow problem. The importance of not satisfying simultaneously the Reynolds and Galilei's number similarity for scaling free surface-flow problems is also discussed. Finally, a scaling parameter was quantified and illustrated for both dimensionless and dimensional form of the free-surface flow (slump-flow) problem.

627307 Influence of Relative Humidity on the Spreadability and Triboelectric Properties of Powders in Additive Manufacturing Processes

Abstract:

Metallic and polymer powders are widely used in Additive Manufacturing (AM) processes, with for example Selective Laser Melting (SLM) and Selective Laser Sintering (SLS). During such operations, successive thin layers of powder are created with a ruler or with a rotating cylinder and then partially sintered with an energy beam. The layer thickness defines the vertical resolution, and the spatial homogeneity of the layer is a key parameter to ensure good and consistent quality of the built parts. Consequently, assessing the spreadability/flowability of powders is required to ensure good and consistent quality of the additively manufactured parts. The flowability of powders depends not only on their intrinsic properties (grain size, shape, etc.) but also on the measurement/processing conditions. Therefore, it is crucial to investigate the influence of the measurement conditions.

In this work, we investigated the influence of the relative humidity on the spreadability, packing dynamics and triboelectric behaviour of several common metallic powders (AlSi7Mg, Inconel, and SS316) used in additive manufacturing applications. The spreadability/flowability of the powders was characterized using the rotating drum method, to reproduce the recoating conditions (flow field and stress state) encountered in an additive manufacturing process. The packing dynamic was measured with the GranuPack instrument to study the bed density. Finally, the triboelectric behaviour of the powders was measured with a GranuCharge.

We show that the relative humidity modifies significantly the cohesiveness, the packing dynamics and the tendency to create electrostatic charges during a flow. The modification of the flow properties and triboelectric behaviour has a direct influence on the spreadability of powders in additive manufacturing applications. This study demonstrates the influence of the environmental conditions on the flow properties of powders used in additive manufacturing applications.

630064 Toward Next Level of Pharmaceutical 3D-Printing through Advanced Lipid-Based Excipients

Abstract:

The opportunities and challenges faced by three-dimensional (3D-) printing in the pharmaceutical field are not based solely on the process itself, but also associated with the material attributes. Until now, polymer-based excipients remain the principal materials employed for dosage formulation and printing of medicine, with very limited trials on lipid-based excipients (LBEs). The high amorphous content of polymers provides them with high flexibility which impacts their 3D printability. LBEs, unlike polymers, are characterized by a crystalline nature which contributes to their brittleness, making their 3D-printing very challenging. Considering the advantages of LBEs compared to polymers, e.g. being naturally occurring materials, predominantly digestible with "Generally Recognized as Safe" (GRAS) status, it is of significant importance to investigate their functionality for 3D-printing of pharmaceutical dosage forms. Recently, polyglycerol esters of fatty acids (PGFAs), also known commercially as Witepsol® PMF, were introduced as the next generation lipid-based excipients offering plenty of potentials for pharmaceutical applications [1]. The broad functionality based on their composition make PGFAs an appealing material for investigation. The synthesis of this group of excipients typically yields a mixture of PGFA as the main component (>80% w/w) and two more species, namely free polyglycerins (PGs) and mono acid fractions (MAF). Tuning the ratio between the three species can influence the material attributes and hence their processability and printability. In this study, we are investigating PGFAs, based on variations in their composition, as an advanced LBE for manufacturing lipid-based drug delivery systems via extrusion-based 3D-printing.

Due to the high number of polyglycerol moieties and free hydroxyl groups per molecule associated with the significant number of hydrogen bonds, and hence higher flexibility, the PG6-C16 partial ester (Witepsol PMF166) was selected as the starting material in this study. Beside the raw material, different blends of PG6-C16 partial esters composed of different ratios of its three components were prepared; either by adding PGs, MAF or both to the raw material. The materials were prepared for extrusion by melting, re-solidifying, crushing and sieving (<800 μm). Filaments of the PG6-C16 partial ester and its blends were manufactured using solid lipid extrusion at temperatures below the melting onset. The extrusion process parameters were adjusted in order to achieve 3D-printable filaments with targeted diameter and ovality using a 1.8 mm-diameter nozzle. By modifying the gear wheels of the printer, the force induced on the filament during 3D-printing was reduced to a minimum, enabling a flawless processability. A drug delivery matrix in the shape of a solid oral tablet ($10\times10\times4$ mm^3) was then printed using the produced filaments. In addition, extrudability and printability of other conventional lipids such as behenoyl polyoxyl-8 glycerides (containing PEG-8 esters) and polyethylene glycol monostearate (containing PEG-32 esters) were compared.

The printability was strongly impacted by the filament flexibility. The flexibility was assessed in terms of the critical bending diameter (CBD), which is the smallest bending diameter at which the material breaks under a bending load. The smaller the CBD, the more flexible is the filament. All LBEs were found extrudable, however, behenoyl polyoxyl-8 glycerides was not processable below its melting onset. The CBD of PG6-C16 partial ester, behenoyl polyoxyl-8 glycerides and polyethylene glycol monostearate were 100, 125, and >223 mm, respectively. The filaments of PG6-C16 partial ester were more flexible, hence easier to handle for printing than other tested lipids. Polyethylene glycol monostearate was unprintable with the present set-up. Varying the composition of PG6-C16 partial ester in terms of PGs and MAF, displayed upgraded performance in extrudability and printability. The highest flexibility was found at the highest ratio of PGs and MAF corresponding to a CBD of 1.8 mm (50 folds lower than that of PG6-C16 partial ester). Higher flexibility led to a more efficient process. The filaments were spoolable during extrusion and less prone to breakage, which, in turn, allowed simultaneous printing of multiple tablets.

In conclusion, PG6-C16 partial ester was effectively extruded and 3D-printed via tuning its composition. This approach is an essential step towards using LBEs as natural and biodegradable alternatives to polymer-based excipients, expanding the pallet of materials that can be used for 3D-printing of medicines. Nevertheless, more opportunities and new functionalities can be enabled through optimizing PGFAs properties.

References

[1] Corzo, C., Lopes, D. G., Lochmann, D., Reyer, S., Stehr, M., & Salar-Behzadi, S. (2020). Novel approach for overcoming the stability challenges of lipid-based excipients. Part 1: Screening of solid-state and physical properties of polyglycerol esters of fatty acids as advanced pharmaceutical excipients. *European journal of pharmaceutics and biopharmaceutics, 148*, 134–147. https://doi.org/10.1016/j.ejpb.2020.01.012

Acknowledgments: The Austrian Funding Agency (FFG)

This work was funded through the Austrian COMET Program by the Austrian Federal Ministry of Transport, Innovation, and Technology (BMVIT), the Austrian Federal Ministry of Economy, Family, and Youth (BMWFJ), and by the State of Styria (Styrian Funding Agency SFG)

627371 CFD Modeling and Liquid Vaporization: Industrial FCC Riser Feed Injection Application

Abstract:

Computational fluid dynamics (CFD) modeling has been used to simulate FCC risers for more than a decade. Riser simulation has been performed with a variety of commercial CFD tools including Barracuda Virtual Reactor.

Historically, for CFD modeling with reactions, the hydrocarbon feed to the FCC riser is assumed to be instantly vaporized, or very simple approximations are made to close the mass balance between phases. However, the liquid injection, liquid penetration, spray angle, droplet size distribution, liquid-solid contacts, vaporization and therefore gas-phase expansion are some of the important parameters that can only be captured if the feed is injected as liquid. Therefore, liquid feed injection, as opposed to injecting the feed in gas phase, is far closer to reality, and critical to typical phenomena of interest for risers.

In this work, we have used Virtual Reactor to simulate an industrial scale FCC riser for a Gulf Coast refinery. The feed is injected to the riser as discrete liquid droplets. Part of the liquid feed directly vaporizes into gas phase while some of the liquid forms a liquid film on the catalyst particles. This film, then vaporizes into gas-phase using the heat from both the gas and the particles. This mechanism is similar to that of real-life liquid feed injection and vaporization in FCC risers.

For this application, the liquid feed injection is characterized by features such as droplet size distribution, spray angle, momentum flux distribution, etc. Liquid penetration, liquid film transfer from droplets to catalyst particles and vaporization are captured with the model. Overall vaporization times obtained from the model agree with commercial riser design criteria. We also demonstrate in this work that liquid feeds, consisting of multiple components with different boiling points, show vaporization profiles agreeing with expectations for such multi-component hydrocarbon feed vaporization.

624268 Machine Learning Based Interaction Force Model for Non-Spherical Particles in Incompressible Flows

Abstract:

Gas solid flow systems are ubiquitous in a variety of applications such as pneumatic transport, combustion of pulverized coal, drying, sand blasting, and combustion of chemical looping systems. Computational fluid dynamics (CFD) has been an effective way to predict the particulate flow and to design the system at a large-scale with the methodologies such as the discrete element method (DEM) and two fluid method. In applications, the interaction force model between gas and non-spherical particles is required, but it is hard to formulate due to the complicated fluid forces which are shape- and orientation-dependent. The objective of this study is to develop the force model for non-spherical particles using the neural network approach which is advantageous to correlate the complicated relationship between variables.

This study uses particle resolved direct numerical simulation (PR-DNS) to collect the datasets for the interaction force such as drag force, lifting force and torque between a particle and an incompressible flow from a low to moderate Reynolds number. The PR-DNS is based on the simplified gas kinetic scheme coupled with the immersed boundary method and the non-spherical particles having different roughness, aspect ratio and orientation are generated by using spherical harmonic function. This approach has the advantage of being more accurate compared to the experimental method because it can reflect the exact shape of the particle and calculate the forces. To develop the force model, we utilize a variational auto-encoder to extract the geometrical features affecting the gas-solid interaction. Furthermore, we apply an artificial neural network to correlate the geometrical features and the flow conditions with the interaction force from the PR-DNS method. The preliminary results using 2,800 datasets show the mean absolute percentage error of 15% for the drag force coefficients, which is lower than 27% when the spherical particles are assumed. This study will provide a drag force model which can be coupled with DEM and consider the high nonlinearity of shape factors, orientation, viscosity, and flow intensity. It will also give a correlation for the lifting force that is important for some cases involving asymmetrical flow around the particles having high aspect ratio. By implementing the force model for non-spherical particles in CFD, this study will improve the accuracy of the multi-phase simulation and utility for a wide range of industrial designs and applications.

627912 Experiments and Simulations of Gas-Solid Flow Dynamics with a Moving Porous Media Model

Abstract:

Continuous spatial particle atomic-layer deposition (ALD) reactors contain billions of fine and ultrafine particles under typical operating conditions. Simulating micron and submicron powders in a reactor with gas dosing zones on the order of centimeters presents a significant modeling challenge. However, experiments and CFD-DEM simulations in MFIX have revealed that the powder bed moves as a solid plug under typical vibratory convection conditions, so we decided to take advantage of the low relative motion between particles by approximating the powder bed as a porous media. A moving porous media model, which utilizes the sliding and layering dynamic mesh capabilities in ANSYS Fluent, was developed to enable efficient simulations of reactor-scale multiphase flow in the continuous vibrating beds used for particle ALD. Treating the packed powder bed as a solid skeleton with an infiltrating gas phase circumvents the computational expense associated with resolving particle-particle collisions. Simulations revealed the effects of packed bed and porous baseplate permeability, vibration intensity, and precursor/purge gas velocity ratios on the flow behavior inside continuous spatial particle ALD reactors. Understanding and predicting the reactor-scale behavior enables process optimization under different reactor conditions.

629430 Optimization of Fluidized Bed Incineration Process for Explosive Waste Treatment Via Artificial Neural Network Surrogate Modeling Method

Abstract:

In this study, the operating conditions for a novel method of explosive waste disposal were optimized to minimize the formation of NOx air emissions and the operating cost. Previous methods of disposal include open burning and detonation and the more recent use of rotary kilns. However, these methods have disadvantages, particularly those associated with NOx emissions. To address this, development of a fluidized bed reactor explosive waste incinerator was proposed. Because this is a new application of the technology, little experimental process data are available. In such cases, computational fluid dynamics (CFD) simulations are typically used to evaluate the preferred operating conditions for the reactor. However, as the number of variables to consider increases, the optimization of the design through CFD simulations alone becomes challenging because of the significant computational time required. Thus, a more efficient method is needed.
Herein, we proposed a surrogate model of CFD output using an artificial neural network (ANN) that can be used to locate the optimal operating conditions. To perform the optimization on seven variables, namely, inlet gas velocity, temperature, pressure, feed particle size, TNT/water mass ratio, filled sand ratio and inlet gas composition, finite ranges of these variables were selected and sampled using a Latin hypercube method, yielding 500 random input samples. CFD simulations were performed using each of these conditions, and the surrogate model was designed to match the results. Using this surrogate model, 50% reduction in NOx emissions or 35% reduction in operating cost from the process was achievable. The surrogate model was then used to perform a sensitivity analysis for each variable, and the inlet gas velocity and reaction temperature had the greatest effect, whereas the TNT/water mass ratio had the smallest effect.
Overall, the fluidized bed reactors show promise with regard to the reduction of NOx emissions and operating cost from explosive waste disposal. In addition, surrogate modeling with ANNs represents a useful method for reducing the CFD computation time used for multi variable optimization of such a fluid bed reactor.

628126 Numerical Analysis and Experimental Verification of Boehmite to Alumina Particles Conversion Under Concentrating Light

Abstract:

Solar energy conversion to chemicals and fuels receives progressively more attention. Many of the possible conversion routes incorporate particles or could be transformed into particle-phase reactions. In particular, particle processes with fine to nano particles (e.g. <250 μm in diameter) diluted into a gas stream, usually a carrier inert gas or a reacting gas mixture, belong to the large family of aerosol processes. Note, that this work does not deal with the other large category of particle solar processes where the particles are of coarse or larger sizes (e.g. > 250 μm in diameter) and of significant mass fractions with reactors in fluidization, moving beds etc. configurations. Most frequently, in aerosol processes the solar energy is delivered to the particles indirectly i.e. by convection via a hot-carrier-gas streams and/ or radiation from the hot walls of the reactor which are heated directly or indirectly by concentrated solar power. Because of the diluted particle-phase, the large distances among the particles, and between the particle flow and the walls diminish the role of conduction.

The literature on aerosol reactors, in drop-tube configurations, heated electrically or indirectly by solar radiation is extensive. There are case studies with these types of reactors producing tungsten carbide, for methane decomposition, the thermochemical reduction of Ceria, and the thermal ZnO dissociation. All previous studies supported that the small particles, in a drop-tube configuration, increased heat and mass transfer rates. At the same time, direct irradiation of the reaction zone and the particle phase potentially can lead to higher solar to chemical energy efficiencies than the indirect solar reactors. Direct irradiation of the aerosol requires an opening or transparent window. Until today, many studies investigated the flow patterns in windowed solar reactors numerically and experimentally, in an effort to address some of the practical challenges e.g. the protection of the reactor opening.

To alleviate some of the aforementioned challenges, this work suggests the use of a transparent tube (e.g. quartz tube) in a drop-tube configuration to directly irradiate the particle flow. Toward this, we developed and validated a single particle model for the conversion of aluminum hydroxides (Boehmite or Gibbsite) and higher Alumina-phases using concentrating light. The two materials have been selected because they are favored for their unique properties in industrial catalysis and refractory ceramics with a very wide market. Next, we investigated, numerically and experimentally, the hypothesis that from a phenomenological point of view the following two processes are equivalent: a) a small quantity of immobile particles (<8 mg) placed within a thin capillary (~ 1mm) exposed to a short pulse of concentrating light, and b) an aerosol of mobile and dilute particles passing through a concentrating light zone with fixed intensity. A hypothesis that was eventually rejected. On the other hand, the combination of the mathematical model and experimental results demonstrated qualitatively that it is possible to achieve the full conversion of Boehmite to γ- or δ-Al_2O_3 in short times with direct irradiation. Sensitivity studies with the mathematical model presented the possibility to fine tune a concentrated light driven aerosol (particle) process to achieve higher selectivity of specific products by taking advantage of the very high heat transfer rates and details of the reaction mechanism. In parallel, the same studies showcased how critical is to have a representative and accurate reaction kinetics model.

624102 Extraction of Particle-Scale Cohesion Parameters from Straightforward Bulk Measurements

Abstract:

Inter-particle cohesion significantly affects the flow behavior of particulate and multiphase systems. Therefore, accurate prediction of inter-particle cohesion is crucial for efficient design and optimization of various industrial unit operations. The state-of-the-art methods to develop cohesion models rely on atomic force microscopy analysis of individual particle surface under specific humidity conditions, which are time consuming, expensive and require specialized training. These methods are not conducive to an industrial setting where feedstock and ambient conditions change frequently. Therefore, exploring possible methods to extract inter-particle cohesion from straightforward bulk measurements is desired. In recent work carried out by Liu et al.[1], defluidization is identified as one example of a bulk behavior that can provide particle-level cohesion. In their work, a simplified square-force cohesion model is proposed, and with the help of discrete element method (DEM) simulations, it is demonstrated that the parameters for the square-force model can be extracted from the defluidization (pressure drop) curve. This defluidization-based approach is restricted to lightly cohesive particles since more cohesive particles result in an erratic defluidization curve. In this work, a similar approach is taken for more cohesive particles – i.e., a system with energy added in order maintain smooth defluidization, namely a rheometer with a rotating geometry, is utilized on particles with mild to moderate levels of cohesion [2]. Additionally, the rheometer can operate with fewer particles, such that corresponding DEM simulations as needed for square-force parameter extraction, is computationally feasible. To accurately extract the inter-particle cohesion parameters, the measured values of the other particle-scale properties (e.g., coefficient of friction, coefficient of restitution) were used as inputs to the DEM simulations. Finally, the same particles were used in a separate experimental system in order to test the validity of square-force model parameters outside of the system from which they were extracted.

[1] P. Liu, C.Q. LaMarche, K.M. Kellogg, C.M. Hrenya, A square-force cohesion model and its extraction from bulk measurements, AIChE Journal. 82 (2018) 1156–11. doi:10.1002/aic.16089.

[2] I. Mishra, P. Liu, A. Shetty, C.M. Hrenya, On the use of a powder rheometer to probe defluidization of cohesive particles, Chem Eng Sci. 214 (2020) 115422. doi:https://doi.org/10.1016/j.ces.2019.115422.

625031 Flame Aerosol Synthesis of Mesoporous Silica for Application in CO_2 Oxidative Dehydrogenation of Propane

Abstract:

Ever since Mobil researchers developed the self-assembled template method for synthesis of the M41S series mesoporous silica in 1992, this class of materials has been applied to numerous applications in catalysis, separation and purification, sensors, and biomaterials based upon its high porosity, ordered pore morphology, and tunable surface groups. Later, researchers developed the aerosol-assisted self-assembly method for the production of mesoporous silica in a process similar to spray drying. Until now, no high porosity mesoporous silica has been produced by flame aerosol technology because the high temperature typical of flame synthesis would destroy the organic template and destroy the pores. Moreover, flame synthesis of silica, *e.g.*, production of fumed silica, generally proceeds by a gas-to-particle conversion route that is inherently incompatible with templating. Here, we report the possibility of producing nanoparticles of mesoporous silica using a unique flame aerosol reactor configuration that separates the flame chemistry and particle formation process into different regions, which favors a much lower reaction temperature. In this process, a liquid solution of precursor and surfactant is injected into the throat of a converging-diverging nozzle, placed downstream of a hydrogen-oxygen flame. The nozzle accelerates the hot combustion gases, which atomize the precursor solution. Solvent evaporation, surfactant self-assembly and silica formation occur in each droplet, during the ~50 ms residence time in the aerosol reactor downstream of the nozzle. The droplets are much smaller and the residence time much shorter than in traditional aerosol-assisted self-assembly processes. With this approach, after calcination to remove the surfactant template, we produced mesoporous silica with a BET surface area of more than 1000 m^2/g, entirely in the form of submicron spheres, with a mean diameter near 100 nm. This approach is also generalizable to other materials, and in our ongoing research we are prepating other mesoporous metal oxides by the same process.

This aerosol synthesis route also allows single-step decoration of the mesoporous silica with another material, which is particularly advantageous for single-step preparation of supported catalysts. Here, as a first example of this, we add chromium nitrate into the silica precursor. As the mesoporous silica forms in droplets, CrO_x nanoparticles are also generated. Upon calcination to remove the templating surfactant, this yields a supported CrO_x catalyst structure. This catalyst shows very promising activity for CO_2 oxidative dehydrogenation of propane, and the flame aerosol technology provides a scalable and continuous process for single-step production of such catalysts. CO_2 oxidative dehydrogenation, in general, which reacts CO_2 with natural gas liquids (ethane, propane, butane) to produce higher value products (ethylene, propylene, butene, and CO) can play an important role in CO_2 utilization.

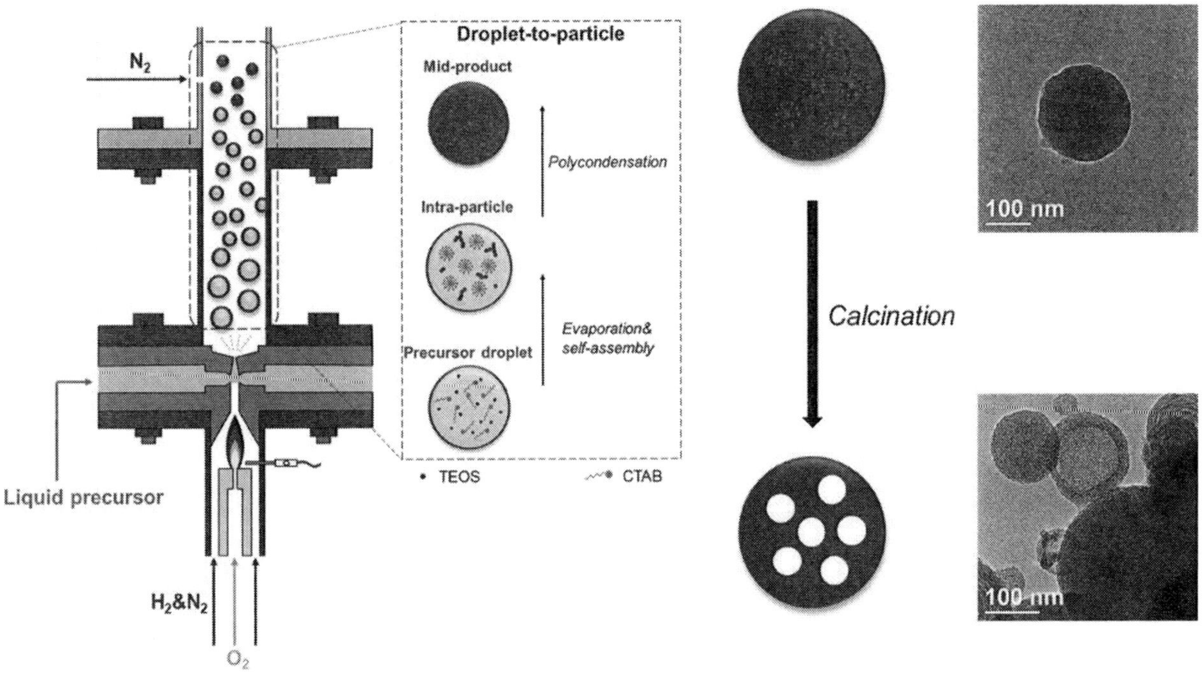
N$_2$
Droplet-to-particle
Mid-product
Polycondensation
Intra-particle
Evaporation&
self-assembly
Precursor droplet
• TEOS
CTAB
Liquid precursor
H$_2$&N$_2$
O$_2$
Calcination
100 nm
100 nm

630005 Role of Water in Methane Oxidation Observed By Transient Drifts at Low Temperatures Using Bimetallic Aupd Catalysts

Abstract:

Methane is an abundant, inexpensive raw material, so it naturally becomes a desirable carbon feedstock for chemical processes. Unfortunately, current approaches are energy-intensive to methane's utilization despite the thermodynamic possibility of low-temperature activation. This awareness beckons for a catalyst that can activate methane at low temperatures. An additional complication is a propensity of overoxidation once one of the C-H bonds is broken. A stabilization agent might mitigate this over-oxidizing tendency.

AuPd nanoparticles have provided some exciting results in methane activation by the Hutchings group in partial oxidation of methane to methanol using AuPd catalysts.[1,2] It is of particular interest to understand and optimize this unique activity of AuPd that the individual components do not possess. In our work, we use *in-situ* vibrational spectroscopy via DRIFTS to observe surface species interactions and the apparent relationship to activity.

We observe changes in surface species as both a function of temperature and composition of the atmosphere both steady-state and transient operating modes. The spectroscopic studies performed on Au, Pd, and AuPd nanoparticles dispersed onto TiO_2 revealed surface hydroxides on the single component catalysts, while in contrast, hydroxide on AuPd is not present in a hydrated atmosphere. Comparing spectroscopic signatures to the steady-state kinetic data, we note that water's presence improves the rate of methane oxidation at low temperatures. At the same time, there is no significant change to activity for the other materials. DRIFTS with a transient composition of feed gas a few interesting data points. Following a purge of dry helium for Au and Pd catalysts, formate remained on the surface to a greater extent than the AuPd bimetallic metal. Observations of spectroscopic signatures and the relationship to activity are addressed, including hydroxide, adsorbed C-H, and carbonyls. In addition to AuPd, a stabilizing material, CuO_x was prepared with AuPd to observe the stabilizing agent's role in the change of material composition. Interestingly, we observed the positive consequence of adsorbed C-H and an improvement in selectivity in the materials with CuO_x.

To summarize, we analyze Au and Pd nanoparticles dispersed on TiO_2 for oxidation of methane by analyzing surface composition via adsorbed species and the relationship to the conversion and the important role of water in this reaction.

(1) Ab Rahim, M. H.; Armstrong, R. D.; Hammond, C.; Dimitratos, N.; Freakley, S. J.; Forde, M. M.; Morgan, D. J.; Lalev, G.; Jenkins, R. L.; Lopez-Sanchez, J. A.; Taylor, S. H.; Hutchings, G. J. Low Temperature Selective Oxidation of Methane to Methanol Using Titania Supported Gold Palladium Copper Catalysts. *Catal. Sci. Technol.* **2016**, *6* (10), 3410–3418. https://doi.org/10.1039/c5cy01586c.

(2) Williams, C.; Carter, J. H.; Dummer, N. F.; Chow, Y. K.; Morgan, D. J.; Yacob, S.; Serna, P.; Willock, D. J.; Meyer, R. J.; Taylor, S. H.; Hutchings, G. J. Selective Oxidation of Methane to Methanol Using Supported AuPd Catalysts Prepared by Stabilizer-Free Sol-Immobilization. *ACS Catal.* **2018**, *8* (3), 2567–2576. https://doi.org/10.1021/acscatal.7b04417.

632199 Continuous Green Millifluidic Synthesis of Five-Fold Palladium Nanorods Using L-Ascorbic Acid and Their Catalytic Application

Abstract:

Pd nanorods (PdNRs) have recently come into attention for their wide array of applications. Sustainable synthesis of five-fold PdNRs have been challenging in the past due to the need to control their aspect ratios very precisely. Millifluidic continuous flow reactors have been explored as mean of maximizing PdNR synthesis by utilizing the ability to precisely control mixing and heat transfer in such reactors. A further development of this idea is to use compartmentalized flow of reaction solution to exploit benefits of improved mixing and heat transfer present in liquid compartments. In an aqueous synthesis of colloidal metal nanostructure such as PdNRs, the final shape taken by a nanocrystal is dictated by the twin structures of seeds and the rates of growth of different crystallographic facets. This study explores the use of L-ascorbic acid as a sustainable reducing agent for sodium tetrachloropalladate (Na_2PdCl_4) precursor in a segmented flow millifluidic reactor with assistant of potassium bromide (KBr) and polyvinylpyrrolidone (PVP) as capping agent and stabilizing agent, respectively. The experiment employs plug flow of reaction solution resulting from vaporization of aqueous solvent at boiling temperature within millifluidic tubing. The effects of various reaction parameters such as reagent concentrations are studied, and these findings are utilized to improve the yield of synthesized PdNRs. The morphologies obtained for the Pd nanocrystals can be icosahedron, octahedron, truncated octahedron, decahedron, rectangular bar, hexagonal and triangular plates, and cube that strongly depends on the synthesis condition and kinetic parameters. The correlation between five-fold PdNR synthesis and the rate of Pd precursor reduction is examined in the kinetic study, and the variation in the reduction rate with reaction progress is investigated to systematically control the shape of Pd nanocrystals. Characterization techniques such as Transmission Electron Microscopy (TEM) and Ultraviolet-Visible (UV-Vis) spectroscopy are used to study the reaction mechanism and kinetics of PdNR synthesis. Palladium nanocrystal is a great catalyst for several reactions, where its activity is strongly dependent on the size, morphology, and different facets expressed on its surface. The ability to precisely control and synthesize Pd nanocrystals with many different shapes including PdNRs provides an outstanding platform to tune their properties for catalytic application and their performance.

624093 Designer Hybrid Colloids: A Study of Gold Adsorption Onto Polystyrene to Control Morphologies of Reactive Nanoparticles

Abstract:

Colloids are commonly used in groundwater remediation as a tool to mitigate the harmful effects of toxic contaminants in complex porous environments, either *via* chemical degradation or removal. Many current technologies seek to implement chemically reactive nanomaterials to degrade harmful contaminants *in situ* and thus prevent them from entering the drinking water supply. However, recent works have shown that many of these nanomaterials are not stable against aggregation, decreasing their reactivity and their resultant ability to treat contaminants effectively. Additional studies have sought to prevent aggregation by applying electrostatic or steric stabilizers on the surfaces of these nanomaterials, which are most commonly catalytic metals, but these solutions do not address a way to easily tune the morphologies, concentrations, or reactivity of injected particles.

We address these shortcomings by developing a framework to control the properties of an inorganic-organic, or hybrid, system comprised of polystyrene colloids that we fabricate *via* Flash NanoPrecipitation (FNP) and smaller gold nanoparticles. We investigate several parameters, including polymer molecular weight, concentration of charged polymer end groups, gold nanoparticle size, and solvent composition, to determine the mechanism of gold adsorption to polystyrene and control particle morphologies. Interestingly, we find that a high concentration of a good solvent must be present during incubation with gold particles in order for adsorption to occur. We show that the presence of a good solvent results in the development of a mobile surface layer on the polymer colloid, allowing for diffusion and adsorption of gold particles within it. Furthermore, we see that as we increase the molecular weight or concentration of charged end groups, we achieve more uniform dispersal of gold nanoparticles on the surfaces of polymer colloids, which not only prevents their aggregation, but creates a larger surface area over which reactions can take place. Finally, we observe internalization of gold nanoparticles within polymer colloids above a threshold concentration of charged end groups, which presents a unique opportunity for catalytic "activation," in which internalized gold particles can be selectively drawn to the surface of polymer colloids. For all obtained morphologies, we test the ability of our hybrid colloids to reduce a model contaminant, p-nitrophenol. Taken together, the results of our work provide a systematic approach to control the morphologies of inorganic-organic hybrid colloids and the unique ability to optimize their reactivity for a given application.

626188 Rheoelectric Characterization of Oxidized Carbon Nanoparticles As Slurry Active Materials

Abstract:

Promising battery systems for renewable energy sources are ones that will provide resilient, flexible power while reducing the carbon emissions, oil demands, and recurring costs of fossil fuel. To this end, the electrochemical flow capacitor (EFC) is an innovative energy storage technology comprised of a slurry active material stored externally from the electrochemical flow cell. This spatial decoupling absent from conventional batteries allows low-cost scaling of EFCs as the price of the overall package asymptotes towards slurry raw material costs. Energy-dense slurries require electrically conductive particles to form a charge-carrying network in an electrolyte solution. Previous work has focused on carbon black because it is inexpensive, has a high surface-area-to-volume ratio, and has excellent electrical conductivity. Unfortunately, carbon black suspensions under flow become undesirably viscous, leading to unacceptable pumping costs for EFC operations. Mitigating high viscosities, however, comes at a trade-off with improving energy storage efficiency as the slurry electrical conductivity and specific capacitance also increase with carbon black content, as described by electrosorption. We hypothesize that by engineering the nanoscale forces of carbon black within suspension, we can design suspension electrodes with better rheological and electrochemical performance. Carbon black particles were functionalized by acid oxidation, presenting a surface chemistry that exhibits a strong, repulsive surface charge and confers colloidal stability. Changes to the size and the surface charge density of the primary agglomerates were derived from dynamic light scattering and zeta potential measurements, respectively. Chemical composition of the surface was obtained through X-ray photoelectron spectroscopy, complemented with Fourier-transform infrared spectroscopy. The active surface was further characterized with porosity and BET surface area measurements through gas sorption. Simultaneous shear stress and impedance measurements of homogenized Carbon black slurries with varying salt concentrations were acquired from a custom-built, cup and bob flow cell. We will develop structure-property relationships for several Carbon black slurries by linking flow-induced microstructural evolutions to observed rheoelectric performances. We anticipate the presence of an optimal position between low viscosity and high conductivity that maximizes the performance to cost ratio of carbon black slurry electrodes.

628780 Utilizing Atomic Layer Deposition to Influence Selectivity for Ni Reverse Water Gas Shift Catalysts

Abstract:

Metal nanoparticle size and dispersion can influence product selectivity of supported metal catalysts. In the case of CO_2 reduction, small, highly dispersed Ni nanoparticles on ceria have shown greater selectivity for the reverse water gas shift (RWGS) reaction over methanation at moderate temperatures. At high Ni loadings, Ni particle size increases, reducing CO selectivity and promoting undesired reactions. In this work, atomic layer deposition (ALD) was leveraged to uniformly deposit Ni nanoparticles on a ceria support with varying Ni loadings. Their performance for RWGS was compared to other Ni/CeO_2 catalysts of equivalent weight loadings synthesized by more traditional techniques.

631590 Targeted Delivery of Surfactants Via Directed Assembly of Nanoparticles at Liquid-Liquid Interfaces By Fine-Tuning Molecular Interactions

Abstract:

A system that combines into a single platform, both controlled assembly and targeted delivery, is demonstrated for the first time. The directed assembly of positively-charged, amine-functionalized polystyrene nanoparticles, PS NPs, at the oil/water interface was accomplished by introducing carboxylic acid functional groups in the oil phase. The electrostatic interactions between the ammonium groups on the NPs and the carboxylic groups on the oil surface was supported by zeta potential measurements and confirmed by laser scanning confocal microscopy. In contrast, negatively-charged, sulfonate-functionalized PS NPs fail to assemble at the interface and remain in the aqueous phase under the same conditions. By tuning the pH or the ionic strength of the medium, we were able to modulate the extent of the interactions between the NPs and the oil surface. After assembly at the oil-water interface, the NPs can release appropriately loaded cargo including surfactant molecules. The released surfactant leads to the formation of a much finer oil-water emulsion. This unique design feature can be exploited in various applications in fields ranging from pharmaceutical and cosmetics to hydrocarbon recovery and oil-spill remediation.

622099 Design and Use of a Thermogelling Methylcellulose Nanoemulsion to Formulate Nanocrystalline Oral Dosage Forms

Abstract:

Oral drug products have become indispensable in modern medicine because of their exceptional patient compliance. However, poor bioavailability of ubiquitous low water-soluble active pharmaceutical ingredients (APIs) and lack of efficient oral drug formulations remain as significant challenges. Nanocrystalline formulations are an attractive route to increase API solubility, but typically require abrasive mechanical milling and several processing steps to create an oral dosage form. Utilizing the dual amphiphilic and thermoresponsive properties of methylcellulose (MC), a new thermogelling nanoemulsion and a facile thermal dripping method are developed for efficient formulation of composite particles with the MC matrix embedded with precisely controlled API nanocrystals. Moreover, a fast and tunable release performance is achieved with the combination of a fast-eroding MC matrix and fast-dissolving API nanocrystals. Using the versatile thermal processing approach, the thermogelling nanoemulsion is easily formulated into a wide variety of dosage forms (nanoparticle suspension, drug tablet, and oral thin film) in a manner that avoid nanomilling. Overall, the proposed thermogelling nanoemulsion platform not only broadens the applications of thermoresponsive nanoemulsions, but also shows great promise for more efficient formulation of oral drug products with high quality and tunable fast release.

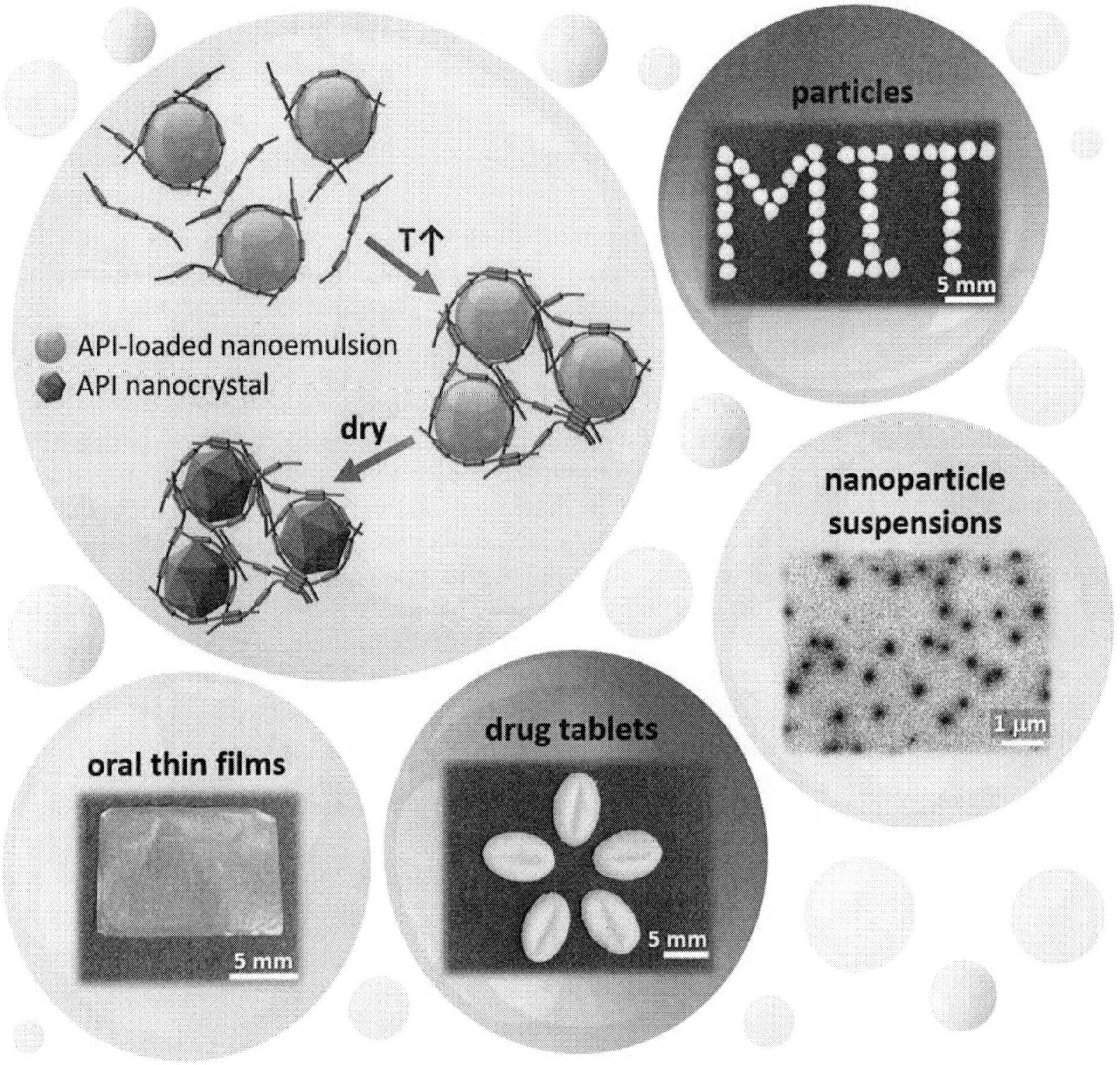
particles
MIT
5 mm
T↑
API-loaded nanoemulsion
API nanocrystal
dry
nanoparticle
suspensions
1 μm
drug tablets
5 mm
oral thin films
5 mm

629913 Particle Formation from the Drying of Liquid Droplets Containing Insoluble Material

Abstract:

Spray-drying is a manufacturing process utilized by a wide range of industries such as food, agriculture, and pharmaceutical for converting a liquid feed containing insoluble solid particles into particulate form. By removing the moisture, it is intended to achieve a final product with desired chemical composition, morphology (e.g., hollow particles, solid cores, agglomerates, etc.), and particle size distribution. There are many variables that affect the drying process, and due to its complexity, traditional trial-and-error methods have been the primary technique in achieving the desired product. The drying conditions and dynamics affect the internal flow patterns within a droplet. The flow patterns are dominated by surface tension and evaporative effects which ultimately disperse the suspended particles in the droplet and finally determine the final product morphology. In the presented work, the drying and internal transport of insoluble particles within a droplet were simulated using a volume of fluid method, where the particle and liquid phases are tightly coupled. In our model, the motions of individual particles are tracked within the droplet, and the impact of particles on the internal flow field is assessed. The model is used to simulate crust formation near the surface of the droplet, where the solid volume fraction increases as droplet evaporates. Simulations are performed at different drying conditions, where temperature, air velocity, and initial droplet diameter are varied to determine their effect of crust formation.

631971 Development of Apap Tablets in a 3D Mold Using Drops of Dissolved Excipients

Abstract:

Excipients are a well-known asset in pharmaceutical tablet formation. Most excipients function as a binder or carrier for an active pharmaceutical ingredients (API) during the granulation process. Some excipients serve multiple roles such as coatings or taste improvement. However, it is important to investigate the individual influences and attributes of excipients prior to investigating process variables. The knowledge derived from such a study enables better formulation design with an overall goal of reducing production costs and improving tablet performance. Natural excipients are useful as the materials are normally hydrophilic and have shown great promise. It is essential to characterize natural excipients for tablet properties to minimize potential issues during processing. Natural excipient will vary in ability and performance. In this study, excipients are in two categories – natural gums and sugars/sweeteners. The performance of these excipients are compared to the performance of two commonly effective and synthetic excipients, polyvinylpyrrolidone (PVP) and polyethylene glycol (PEG). Likewise, the compatibility of the API and excipient should be investigated. The objective of this research is to determine the feasibility of adopting select natural gums and sweeteners as safe and economic replacement excipients based on characterization and performance testing.

Tablet formation occurred in molds that were created using a Flash-Forge 3D printer. Paracetamol (APAP) in powder form was chosen as the API for this study. Dry APAP powder and liquid binder solution at various saturation levels were placed into holes created on each mold. The liquid binder solutions were created by dissolving the powder excipient into deionized water. The molds were placed in a drying oven for 24 hours. The weight of each granule was recorded with size and shape determined using image analysis software. Tablet performance was determined using friability testing, compression testing, dissolution testing, and disintegration testing. Further analysis included X-ray diffraction and ATR FTIR.

The purpose of using a 3D printed mold was to control the overall shape of the tablets and to ensure each tablet contain the same amount of APAP. Various mold designs were investigated to reduce moisture in the tablets and increase heat transfer within the molds. The designs improved the previous method of using a powder APAP bed. Overall, the type of excipient influenced tablet performance during each testing method. The saturation level of the liquid binder affected both friability and disintegration. The new 3D-mold method produced tablets of common size, shape, and weight which was an improvement from previous research. The compression testing showed that the synthetic excipients are able to withstand a larger stress than the natural excipients. Rapid dissolution kinetics (greater than 60% dissolution in 15 minutes) were noted for most of the excipients investigated. During friability testing, tablets made from pectin were completely destroyed while the other excipients had less than 10% friability. The results also showed that increasing the saturation level of the liquid binder solution will cause a decrease in the percent friability. Disintegration time varied with respect to the different excipients. Based on this study, natural excipients such as d-sorbitol have the potential to replace synthetic binders as pharmaceutical excipients.

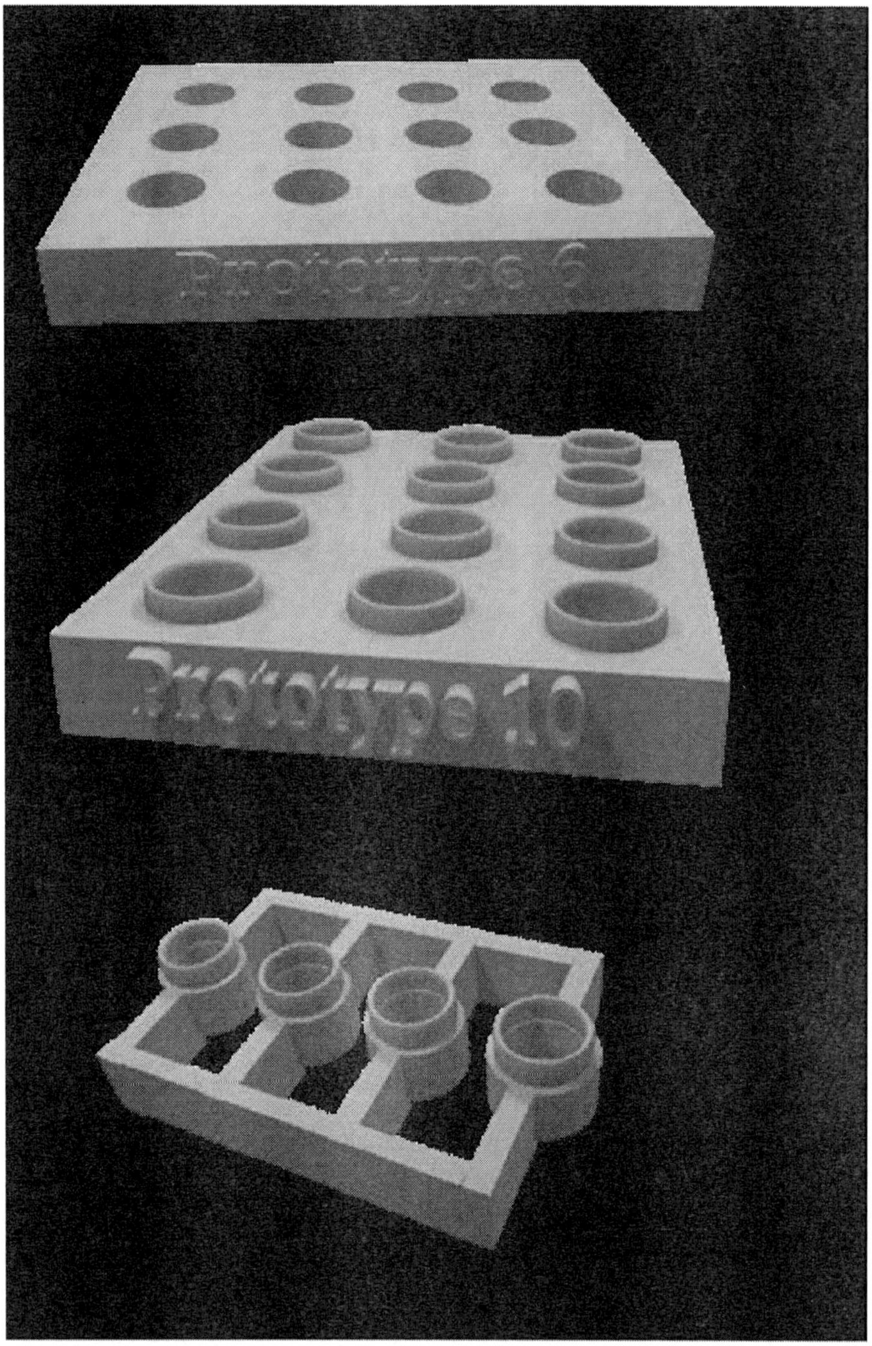
Prototype 6
Prototype 10

632979 Wet Milling with in-Line Fines Dissolution

Abstract:

Solid-state properties, such as particle size distribution (PSD) and morphology, affect downstream processing steps as well as critical quality attributes (CQAs) of the drug substance. Several post-processing techniques exist for the size reduction and impacting the PSD. Two of the most commonly used approaches for crystal breakage are milling and sonication.

In this study, a wet mill is coupled with a heat exchanger in a closed-loop system to dissolve generated fines from the wet mill and to obtain the desired particle size distribution. Coupling the dissolution step with the breakage step minimizes the time and cost of post-processing, yet it requires a detailed understanding of the system. In addition to the heat exchanger, frictional forces between the crystals and the wet mill also increase the slurry temperature. The heat exchanger design is optimized for the desired PSD using a mathematical model for simultaneous breakage and dissolution using gPROMS from gFormulated products. Initially, breakage and dissolution are studied separately to understand each of these mechanisms and evaluate the necessary coefficients for breakage kernel and dissolution rate constants. A series of milling and dissolution experiments are performed on Carbamazepine (CBZ), and the final PSDs are imported into the model to validate the model parameters. Lastly, the obtained kernels are used to construct a coupled wet mill and heat exchanger model and the model is evaluated against experimental datasets. This model would be used to optimize the heat exchanger design to obtain desired PSD from a given wet milling process.

MK, MB, DP, JG NN are employees or a retiree of AbbVie and may own AbbVie stock. PC was an AbbVie intern and is currently a doctoral student at UIC with no additional conflicts to disclose. MS is a Professor at UIC with no additional conflicts to disclose. AbbVie sponsored and funded the study; contributed to the design; participated in the collection, analysis, and interpretation of data, and in writing, reviewing, and approval of the final publication.

631469 Autocatalytic Initiation Followed By Oriented Attachment Governs the Nucleation and Crystal Growth of MOF Crystals

Abstract:

Metal-Organic Frameworks (MOFs) have gained significant importance in the field of catalysis, membrane separations, adsorption, and drug delivery. The wide spectrum of the applicability of MOFs arises due to the underlying crystal structure. In such crystal structures, a metal node forms a template for the attachment of the organic linker resulting in the formation of an infinite network. The crystal structure is further stabilized due to the reorientation of the crystalline network yielding highly symmetric and porous crystalline structures. However, the limited understanding of a large number of reactions involved and the subsequent growth of MOF crystals prevents effective control and rapid synthesis of the MOFs. Furthermore, it is observed that MOF reactions are surface stabilized and follow a non-classical nucleation mechanism. To achieve greater control on the crystallization of MOF and to design better strategies for synthesizing theoretically predicted MOFs, it is necessary to obtain the quantitative contribution of all the reactions which occur during the synthesis. In this work, we present a simple microkinetic approach requiring minimal computational power to obtain rate contribution of initiation, chain growth, step growth, and termination reactions. The approach also validates that MOF initiation follows an autocatalytic mechanism and reproduces experimental crystalline volume fraction and grain size dynamics obtained from SAXS/WAXS studies of UiO-66 synthesis. The microkinetic rate equations effectively capture the rate-limiting step of the MOF synthesis and allow calculations of nucleation and growth rates as a function of reaction extent. Results suggest that the autocatalytic initiation followed by oriented attachment governs the nucleation and growth of MOF crystals.

628645 Controlling Polymorphism and Orientation of Nu-901/Nu-1000 Metal–Organic Framework Thin Films

Abstract:

NU-1000, a zirconium (Zr)-based metal–organic framework (MOF), is a promising candidate for heterogeneous catalysis, gas storage, electrocatalysis, and drug-delivery applications due to its large pore size and mesoporous structure. However, the synthesis of NU-1000 may produce another polymorph NU-901, which has a smaller average pore size and pore volume than NU-1000. Similarly, the presence of NU-1000 as a phase impurity in NU-901 crystallites is undesired. Although phase-pure NU-901 and NU-1000 have been successfully synthesized in bulk, multiple applications such as electrocatalysis and separation membranes require the formation of thin films. In this study, we utilize self-assembled monolayers and crystal engineering to control the polymorphism and orientation of NU-901/NU-1000 thin films. We report the fabrication of thin films of NU-901 and NU-1000 via a solvothermal method by functionalizing the substrate with carboxyl (−COOH) tail groups. This synthesis produces phase-pure hexagonal rod-shaped NU-1000 crystals and nearly phase-pure prolate-shaped NU-901 crystal as revealed by scanning electron microscope (SEM), powder X-ray diffraction (PXRD), and nitrogen adsorption isotherm analyses. Furthermore, we control the orientation of NU-1000 crystallites on the fluorine-doped tin oxide (FTO) substrate by controlling the nucleation density of the MOFs on the substrate. We hypothesize that heating the functionalized substrate in a Zr-oxo cluster solution preceding solvothermal synthesis results in the coordination of Zr-oxo clusters to the (−COOH) groups of the substrate, which promotes a higher nucleation density of NU-1000 on the substrate, resulting in the perpendicular growth of NU-1000 during crystal formation. The method developed to alter the orientation of NU-1000 may pave the way for controlling the orientation of other anisotropic MOFs, which has implications in catalysis, separation, and drug delivery.

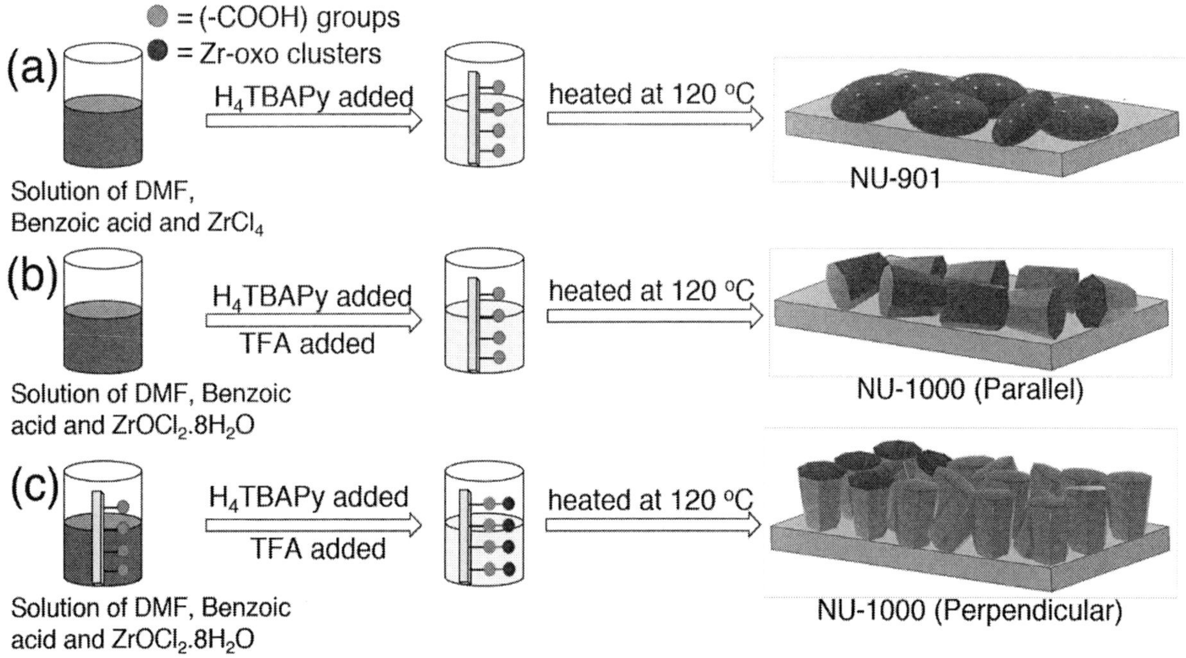

632517 Thermodynamic Modeling of Competing Crystal Species from a MIL-53 Metal Organic Framework (MOF) Reaction

Abstract:

The chemical speciation of a MIL-53 MOF reaction as a function of a broad range of temperatures from 350 to 493 K and different reactant concentrations is estimated by numerically solving a set of coupled equilibrium and conservation equations with nonlinear and linear characteristics, respectively, and propagating the numerical solver on such set through a changing value of temperatures and/or reactant concentrations. Chemical speciation of the reaction was also expressed graphically as three-dimensional surfaces with the intent to map critical synthesis regimes which would optimize the crystallization of MOF and reduce that of competing species.

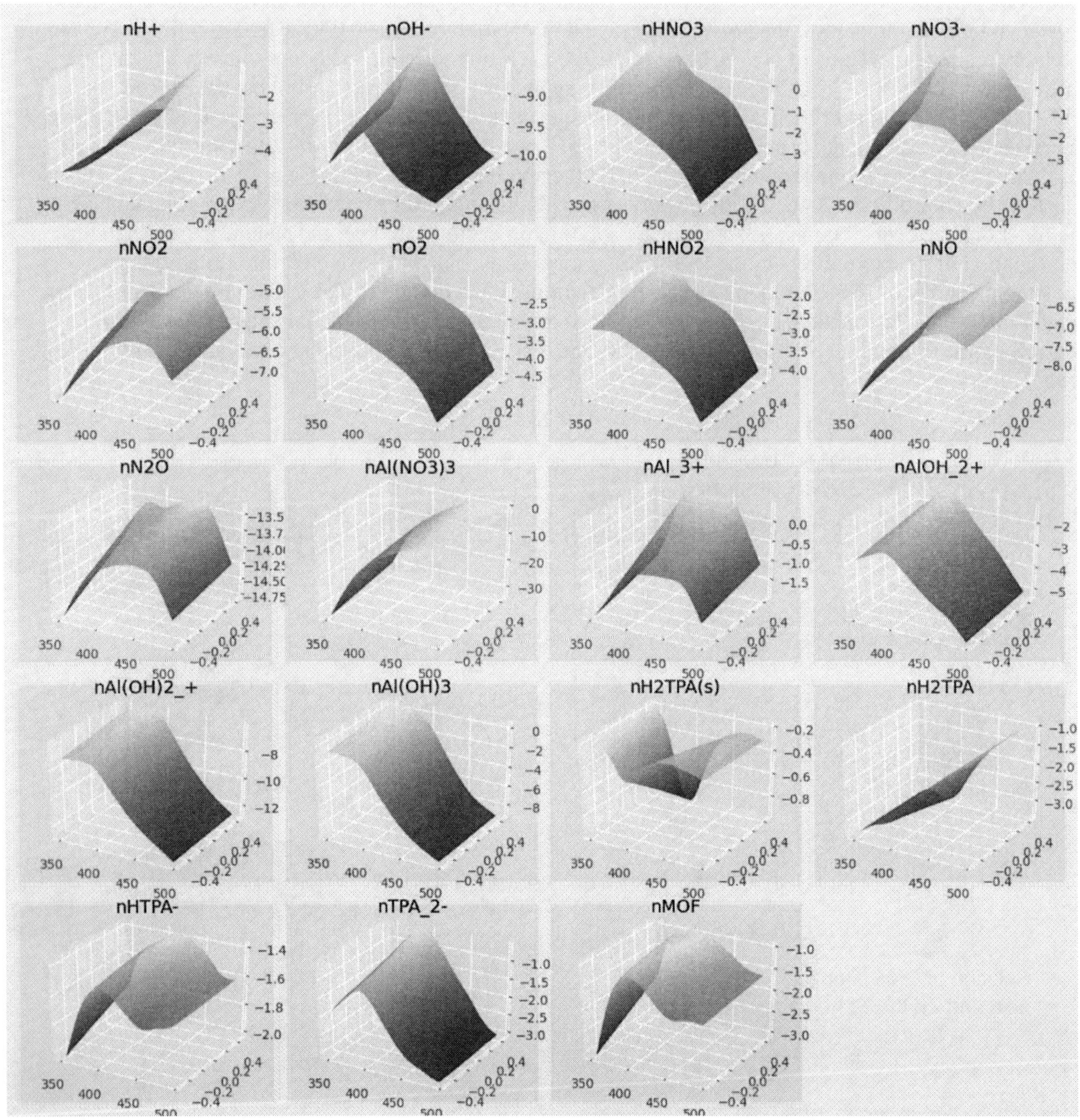

628374 Preparation and Evaluation of Hafnium Oxide Nanoparticle CT Contrast Agents

Abstract:

The high atomic number (Z = 72), high electron density, and chemical stability of hafnium oxide (hafnia, HfO_2) makes its nanoparticles good candidates for biomedical applications as x-ray contrast agents in computed tomography (CT) and as radiotherapy enhancers. Currently, iodine x-ray contrast agents are commonly used clinically. However, studies suggest that hafnia has better overall performance compared to iodine at clinical CT tube potentials. In addition, iodine-based CT contrast agents show increase in retention in the kidneys and can leads to low tolerance to patients with chronic kidney disease. Gold nanoparticles are also of interest as x-ray contrast agents and radiotherapy enhancers. Hafnia nanoparticles could be a lower-cost alternative to gold nanoparticles ($8/g Hf vs. $56/g Au).

Previous studies describing preparation of hafnia nanoparticles for use as CT contrast agents used particles synthesized using sol-gel and direct precipitation methods. However, these approaches have limited size and shape control and can result in contamination leading to cytotoxicity. Although thermal decomposition synthesis methods could provide for improved size control, they remain poorly investigated for synthesis of hafnia nanoparticles. Additionally, thermal decomposition methods result in particles coated with an organic layer and require phase transfer to aqueous media to enable biomedical applications.

Flash nanoprecipitation (FNP) is a simple and scalable method to formulate nanoparticles consisting of a hydrophobic core and an outer hydrophilic layer through rapid mixing and diffusion limited aggregation of hydrophobic compounds and particles and amphiphilic block copolymers. FNP has been widely used to formulate organic drug carriers and, to a lesser extent, composite nanoparticles consisting of inorganic nanocrystals and co-core organic excipients such as fluorophores and drugs.

Here, we report the preparation of hafnia nanoparticle clusters (HNCs) using hafnia nanoparticles synthesized through thermal decomposition of hafnium trifluoroacetate in oleylamine. Hafnia nanoparticles and hydrophobic green fluorophore (3,3′-Dioctadecyloxacarbocyanine perchlorate) with polylactic acid (PLA) as co-core polymer were encapsulated in poly(lactic acid)-b-poly(ethylene glycol) (PLA-b-PEG) by FNP using a new 3D printed multi-inlet vortex mixer. HNCs of hydrodynamic size ranging from 100 to 200 nm were prepared by adjusting FNP conditions. The CT contrast performance of HNCs was compared to the clinical CT contrast agent OMNIPAQUE (active ingredient, iohexol), suggesting better CT contrast for the HNCs. The HNCs were found to be colloidally stable in cell culture media, with negligible change in hydrodynamic diameter over a period of 24 hours. Results for in vitro cytotoxicity and pharmacokinetics of the HNCs in mice will also be presented. This study will be the first to investigate the CT performance, cytotoxicity, and pharmacokinetic of hafnia nanoparticles synthesized from thermal decomposition. We aim to find an excellent candidate to substitute current blood pool imaging agents for CT with lower cost, better biocompatibility, and higher tolerance to patients with chronic kidney disease.

629743 Encapsulated Platinum Nanoparticles, Targeted Therapy for Triple Negative Breast Cancer

Abstract:

Nanotechnology is a multidisciplinary field that has many different applications including cancer treatment and detection. Noble metal nanoparticles such as gold, silver or palladium, have shown major potential in the field of medicine and pharmaceutics. In particular, platinum nanoparticles (PtNPs) have been reported with promising anticancer results. The activity of nanoparticles is based on small size and high surface area; these properties allow them to penetrate biofilms as well as influence intracellular mechanisms. Recent studies indicate that Pt NPs can be used as a therapy for cancer showing a limited toxicity to healthy cells. Hence, in this work, we developed a new treatment option for TNBC based on Pt NPs. Moreover, to enhance the circulation at the tumor site, a delivery system was proposed. Poly(lactide-co-glycolide) (PLGA) particles were synthesized as a delivery system for PtNPs. PLGA is a copolymer of poly(lactic acid) (PLA) and poly(glycolic acid) (PGA), two monomers that are metabolized by the body, making PLGA toxicity minimal. The encapsulation of PtNPs within PLGA nanoparticles was performed by nanoemulsion.

Nanoparticles that circulate in the bloodstream are recognized by proteins. Proteins attach to the surface forming a nanoparticle-protein complex (protein corona) that leads to phagocytic system recognition and removal from the bloodstream. To enhance the circulation life-time of the particles, surface modifications with some polymers such as polyethylene glycol (PEG) can be performed. Surface modifications with PEG is a commonly used procedure to enhance the properties of drug systems. Active targeting which includes antibodies attached to the surface provides the most effective therapy. Several targeted therapies can be used such as ligands of specific biomarkers, monoclonal antibodies, peptides, and aptamers. Triple negative breast cancer is defined by the lack of overexpression of the three common receptors used to classify breast cancer. Therefore, there are very few available targets. Epidermal growth factor (EGF) is a receptor that promotes cell proliferation migration and angiogenesis (development of new blood vessels). EGFR is overexpressed in many tumors including breast cancer. Therefore, many strategies based on antibodies have been studied to block EGFR. The anti-EGFR monoclonal antibody, Cetuximab, has shown promise for targeting metastatic TNBC.

In the present study, PtNPs were synthesized and the anticancer activity was tested against TNBC, showing a remarkable cell death. As the major drawback of traditional chemotherapy is the toxicity, cell viability experiments were also performed on fibroblasts. The results showed no apparent toxicity of Pt NPs for healthy cells in contrast to cisplatin (a common chemotherapeutic drug). Encapsulation of PtNPs within PLGA was performed by nanoemlusion with a 60% loading percentage. Surface modifications were performed to prevent protein corona formation and achieve a higher anticancer effect. Cytotoxicity with TNBC of PtNPs encapsulated in PLGA with active and passive targeting were compared, demonstrating a higher cell death by particles conjugated with the antibody. In summary, this work provides evidence of a potential new treatment for TNBC based on Pt NPs with no apparent toxicity for healthy cells, overcoming one of the major drawbacks of current treatments.

630376 Engineering High-Throughput Gold Nanoshell-Liposomes for Effective mRNA Delivery

Abstract:

Altering immune system cells such as T-cells or NK-cells with specific protein targeting moieties is a promising route for cancer and infectious disease treatment[1]. Non-viral vector delivery of mRNA to code for these proteins is advantageous due to the relative ease of mRNA manufacture and complete absence of genomic integration[2]. mRNA only needs to reach the cell cytoplasm to catalyze the production of exogenous proteins. mRNA is transiently active and does not integrate into the genome, so cell modifications are not permanent or replicated, which makes it safer than DNA delivered by viral vectors[3]. However, efficient intracellular delivery of mRNA to the cell cytosol in vitro or in vivo continues to be a major hurdle. The large molecular weight (10^5–10^6 Da) and high negative charge of mRNA prevent mRNA from crossing cellular membranes[4]. Moreover, mRNA is an inherently unstable molecule, which is highly prone to degradation by 5' exonucleases, 3' exonucleases, and endonucleases[5]. Consequently, development of high throughput and efficient delivery systems are imperative for manufacturing protein-modified immune cells ex vivo. Different strategies have been investigated to improve RNA delivery, such as condensation of mRNA in lipid nanoparticles as in the Moderna and Pfizer vaccines or electroporation, both of which are inefficient and often toxic. There are few available viral vectors for mRNA, and all have drawbacks associated with genome integration, and possible host rejection (immunogenicity and cytotoxicity)[6].

To create a high throughput, high viability ex-vivo mRNA delivery we have encapsulated mRNA in neutral liposomes that are then coupled to plasmon-resonant hollow gold nanoshells (HGN). The HGN plasmon resonance can be tuned to absorb physiologically friendly near infra-red light of 650 – 1040 nm wavelength[7]. After mixing the HGN-liposomes with the cells to be transfected, the mixture can be irradiated with picosecond pulsed near infra-red light that induce vapor nanobubbles around the HGN. These nanobubbles grow and collapse, thereby rupturing both the cell and liposome membrane, thereby releasing the mRNA to the cell cytoplasm. Once delivered, the mRNA can produce the proteins of interest (Fig 1). Here we present an extensive analysis into HGN synthesis, sterilization and liposome coupling to minimize mRNA degradation, greatly improving the transfection efficiency. We show high efficiency mRNA-induced green fluorescent protein production in Jurkat (non-adherent) and HEK-273 (adherent) cells.

Synthesis and characterization: HGN were synthesized using the galvanic replacement reaction from silver nanoparticle templates in aqueous solution. The silver template size and the gold salt to silver ratio determines the size and shell thickness of the HGN, which in turn determines the plasmon resonance wavelength. The HGN were characterized using UV-Vis, and Nanosight particle counting and TEM imaging and yielded an average size of 30 nm and shell thicknesses of 2- 5 nm giving plasmon resonances of 650 -1040 nm, depending on the shell diameter to thickness ratio. Dipalmitoylphosphatidylcholine (DPPC) with small mole fraction of thiol terminated distearoylphosphatidylethanolamine- polyethylene glycol formed liposomes using the thin film hydration method, followed by manual extrusion to 200 nm. These liposomes were characterized by Nanosight and TEM. Different batches of liposomes were made containing carboxyfluorescin (CF), quantum dots or eGFP-mRNA. The liposomes were, thereafter, coated with gold nanoparticles by mixing the HGN with the liposomes; the HGN reacted with the thiol terminated PEG such that 1-3 HGN were bound to each liposome.

In-vitro testing: Transfection was achieved by suspension of the HGN-liposomal mixture in cell media at different concentrations. The HGN in cells were triggered by picosecond pulses of near-infrared light to release dye and/or mRNA from liposomes to cell. Cytotoxicity was measured using an MTT assay kit. The uptake of HGN-liposome was confirmed and quantified using flow cytometry and fluorescent microscope images. Transfection efficiency of the system was measured for all the GFP-positive formulations.

Results: Our results showcase successful HGN-liposome synthesis and characterization. The HGN are tethered to the liposomes as shown by TEM. Intracellular uptake of the HGN-Liposome contents was successful demonstrated by fluorescent microscopy and flow cytometry of 6-carboxyfluorescein (small molecule), quantum dots (2 - 5 nm in diameter) and expression of eGFP. The liposome-HGN showed little to no toxicity to both adherent and non-adherent cells. NIR pico-second light mediated release showed efficient uptake of carboxyfluorescein, quantum dots and eGFP mRNA. The HGN/mRNA mixture had successful cell transfection with high efficiency of eGPF expression with no minimal effects on cell viability post transfection. We plan to improve the throughput rate by increasing the flow rate of the cells through the laser light pulses to achieve a highly efficient method of ex-vivo cell modification.

References:

1. Ben-Shmuel, A., Biber, G. and Barda-Saad, M., 2020. Unleashing Natural Killer Cells in the Tumor Microenvironment–The Next Generation of Immunotherapy?. Frontiers in immunology, 11, p.275.
2. Sahin, U., Karikó, K. and Türeci, Ö., 2014. mRNA-based therapeutics—developing a new class of drugs. Nature reviews Drug discovery, 13(10), pp.759-780.
3. Warren, L., Manos, P.D., Ahfeldt, T., Loh, Y.H., Li, H., Lau, F., Ebina, W., Mandal, P.K., Smith, Z.D., Meissner, A. and Daley, G.Q., 2010. Highly efficient reprogramming to pluripotency and directed differentiation of human cells with synthetic modified mRNA. Cell stem cell, 7(5), pp.618-630.
4. Yin, H., Kanasty, R.L., Eltoukhy, A.A., Vegas, A.J., Dorkin, J.R. and Anderson, D.G., 2014. Non-viral vectors for gene-based therapy. Nature Reviews Genetics, 15(8), pp.541-555.
5. Mitchell, P. and Tollervey, D., 2000. mRNA stability in eukaryotes. Current opinion in genetics & development, 10(2), pp.193-198.
6. Weng, Y., Li, C., Yang, T., Hu, B., Zhang, M., Guo, S., Xiao, H., Liang, X.J. and Huang, Y., 2020. The challenge and prospect of mRNA therapeutics landscape. Biotechnology advances, 40, p.107534.
7. Ogunyankin, M.O., Shin, J.E., Lapotko, D.O., Ferry, V.E. and Zasadzinski, J.A., 2018. Optimizing the NIR fluence threshold for nanobubble generation by controlled synthesis of 10–40 nm hollow gold nanoshells. Advanced functional materials, 28(10), p.1705272.

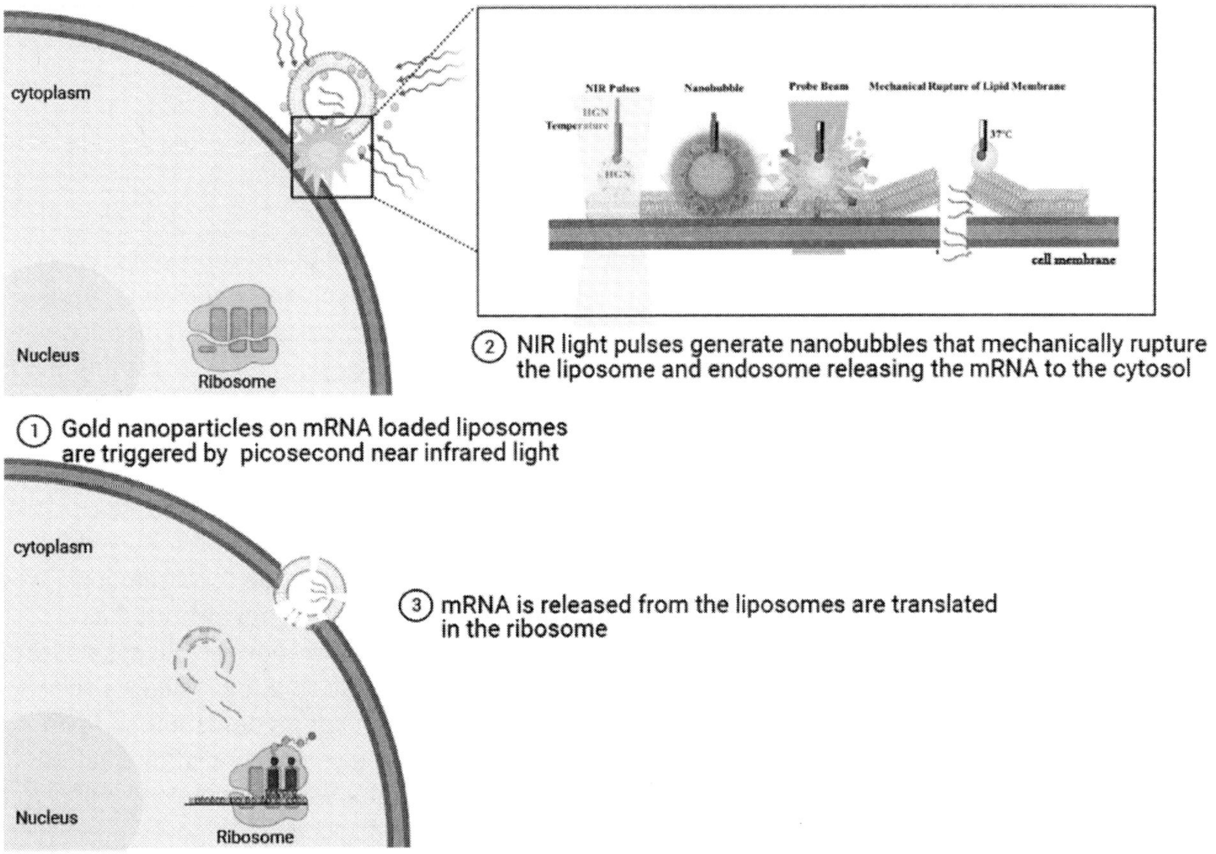
cytoplasm
NIR Pulses
Nanobubble
Probe Beam
Mechanical Rupture of Lipid Membrane
HGN
Temperature
HGN
37°C
cell membrane
Nucleus
Ribosome
② NIR light pulses generate nanobubbles that mechanically rupture the liposome and endosome releasing the mRNA to the cytosol
① Gold nanoparticles on mRNA loaded liposomes are triggered by picosecond near infrared light
cytoplasm
③ mRNA is released from the liposomes are translated in the ribosome
Nucleus
Ribosome

630563 Fabricating Gold Nanoparticles-Loaded Polyvinyl Alcohol Contact Lens for Laser Protection

Abstract:

Background/motivation: Gold nanoparticle-loaded contact lenses could be useful for many ophthalmic applications, such as laser protection, color blindness management, ophthalmic drug delivery, etc. Gold nanoparticles are normally incorporated into contact lenses by polymerization of contact lens monomers mixed with gold nanoparticles. However, achieving high gold loading in the lens and maintaining gold structure are the major challenges because of the compatibility issue between gold nanoparticles and the formation used for contact lens. Here, we report a method for fabricating polyvinyl alcohol (PVA) contact lens with the gold nanoparticles incorporation of high loading for ocular laser protection.

Methods: Gold nanoparticles are synthesized via the reduction of gold chloride by citrate ions, which is known as Turkevich method. PVA is not only used as contact lens matrix, but also used as stabilizer during the synthesis. By adding PVA, gold nanoparticles are also more stable against the salt in the solution, which enables a greater yield of gold nanoparticles during the synthesis. The concentration of gold nanoparticles can be further adjusted by anti-solvent precipitation and particle re-dispersion. Gold nanoparticles separate into the PVA phase due to the non-bonded interaction between gold and PVA so that the phase separation achieved by adding anti-solvent contributes to the collection of Au/PVA nanocomposite. The nanocomposite can be redispersed in a good solvent without affecting the morphology of gold nanoparticles. Au/PVA contact lens is fabricated by physical crosslinking method through cyclic freezing/thawing process. When the freezing/thawing process is repeated, more PVA crystallites will be generated, which serve as crosslinkers in the PVA matrix.

Results: SEM image shows that gold nanoparticles are uniformly distributed in the contact lens with no aggregates observed. From the transmittance measurement, gold nanoparticles in the contact lens absorb wavelengths of around 530 nm due to the localized surface plasmon resonance (LSPR), which is used for laser protective application. When increasing the gold loading, more shielding effect will be achieved. The power of the laser is attenuated by more than 60% when the gold loading reaches 4% (wt%) and more than 90% shielding can be achieved when the gold loading reaches 7% (wt%). Under the laser radiation heat can be generated from the lens, which can also be utilized for Meibomian dry-eye treatment.

Conclusions: We developed a new method for gold nanoparticle PVA-based contact lens fabrication. PVA serves as both stabilizer during the particle synthesis and the contact lens matrix. High gold loading in contact lens can be achieved by concentrating gold nanoparticles through anti-solvent precipitation while maintaining the original morphology of the particles. Gold nanoparticles in the contact lens absorb the light resulting a shielding effect against the laser radiation.

630628 Gold Nanoparticle Synthesis in Contact Lenses for Drug-Less Ocular Cystinosis Treatment

Abstract:

Background/motivation: Ocular cystinosis is an orphan disease characterized by the accumulation of cystine crystals in the stroma layers of the cornea. Currently, ocular cystinosis are commonly treated by installing cysteamine eye drops for cystine clearance. However, cysteamine is prone to oxidation at body temperature before diffusing through the cornea, which significantly limit its bioavailability. Here, we develop a novel contact lens with immobilized and stable gold nanoparticles, which can react with cystine molecules in the tear film. Therefore, instead of releasing drug to clear the cystine crystals inside the cornea, cystine molecules in the tear film diffused into the contact lens and attached to the surface of gold nanoparticles, which will reduce the cystine concentration in the tear film and lead to a dissolution of cystine in the cornea, eventually.

Methods: Surfactants might affect the binding of cystine to the gold. Therefore, to avoid the surfactant, gold nanoparticles are synthesized inside the preformed commercial contact lenses (Acuvue TruEye and Acuvue Moist contact lenses) by first loading the lenses with gold precursor followed by reduction to the gold atoms, which will then nucleate and grow to gold nanoparticles inside the contact lens. The gold loading is adjusted by either increase the concentration of gold precursor solution or repeating the manufacturing process. The gold nanoparticles are characterized by SEM, XRD, UV-Vis spectroscopy and the gold loading in the contact lens is determined by measuring the mass change of the dehydrate lenses. The mechanism of the synthesis is revealed by investigating the transport of the gold precursor and reducing agents. The amount of the reacted cystine is characterized by monitoring the concentration of cystine solution, after soaking the gold nanoparticle-loaded contact lenses in cystine solution.

Results: The gold loading in the contact lens increases linearly with the gold precursor concentration and the number of repeating manufacturing cycles. Results also show that stronger reducing agent leads to a lower gold loss during the synthesis, and the gold loading is higher in the Moist contact lens due to higher diffusivity in the lens. When repeating the manufacturing process, smaller gold nanoparticles of 10 nm (vs 100 nm for Moist) in size are generated inside the TruEye contact lens, which results in a higher cystine binding of around 47 microgram/lens in 5 hours (vs 33 microgram/lens in 2 hours for Moist) due to the higher surface area.

Conclusion: A method of gold nanoparticle contact lens manufacturing is reported, which is used for drugless ocular cystinosis treatment by binding significant amount cystine on the surface of gold nanoparticles inside the contact lens. Moist lens has its benefit for the rate of cystine removal, while TruEye lens has its advantage for cystine binding amount.

632206 Modulation of Neutrophil Extracellular Trap Formation through Polymer Coating of Metal Oxide Nanoparticles

Abstract:

Given that 1 in 8 women develop breast cancer throughout their lifetime, the disease is a substantial public health burden. Patient morbidity and mortality occurs primarily due to metastatic spread, but is compounded by thrombosis, or the formation of blood clots. Recently, neutrophils activated by the inflammatory tumor microenvironment have been observed to release web-like DNA fibers known as neutrophil extracellular traps (NETs). NETs have been linked to several negative downstream effects including thrombosis, damage to vessel endothelial cells, and capture of circulating tumor cells to promote metastasis. Due to the pathogenic role of NETs in breast cancer, it is critical to ensure that contrast agents used in diagnostic imaging do not elicit this response. Breast MRI is often used as a supplement to mammography for high-risk women, as it detects breast cancers that may be missed by mammography and provides superior soft-tissue contrast; however, the standard MRI contrast agents used in the clinic (e.g. gadolinium chelates) cause high false-positive rates, as high as 25%. Thus, alternative contrast agents to improve diagnostic accuracy are being developed, including metal oxide nanoparticles such as iron oxide (Fe_3O_4) and manganese oxide (MnO). Prior studies have demonstrated that certain nanoparticle formulations may promote NETosis, which could encourage cancer progression. The goal of this study was to evaluate the effect of altered polymer coating of MnO nanoparticles on NET formation.

MnO nanocrystals were synthesized using thermal decomposition of Mn (II) acetylacetonate in dibenzyl ether and oleylamine at 280°C. Nanocrystal size and chemistry were evaluated with TEM, XRD, and FTIR. Hydrophobic MnO nanocrystals were encapsulated with biodegradable poly(lactic-co-glycolic acid) (PLGA) and increasing amounts of polyethylene glycol (PEG) including 0%, 2.5%, 5%, and 10%. Hydrodynamic particle size was evaluated by DLS and confirmed to be ~180 nm. Neutrophils were isolated from the bone marrow of healthy female BALB/c mice femurs via density gradient. Neutrophils were stained with CellTracker Deep Red, plated, and stimulated for 3 hours with either media only (negative control), phorbol-12-myristate-13-acetate (PMA) (positive control), unencapsulated MnO nanoparticles, or PLGA MnO nanoparticles with 0-10% attached PEG. DNA was stained with Hoechst and plates were imaged with an inverted confocal fluorescence microscope to evaluate total NET formation. Nikon General Analysis software was used to quantify the number of neutrophils and NETs per field of view.

The greatest NET formation resulted from PMA stimulated neutrophils with an average of 35.83 NETs per 100 neutrophils. Unstimulated neutrophils caused the second-highest amount of NET formation at 19.28 NETs per 100 neutrophils. Bare MnO nanocrystals and PLGA MnO nanoparticles without PEG attached resulted in similar NETosis to unstimulated controls, whereas particles coated with a low percentage of PEG, provoked less NETosis than did stimulated neutrophils; 2.5% PEG PLGA nanoparticles had the smallest response of 11.06 NETs per 100 neutrophils. These results indicated that polymer-coated MnO nanoparticles do not enhance NETosis, supporting their safety as future breast cancer MRI contrast agents. Future studies will elucidate the mechanism responsible for the observed reduced NETotic response with nanoparticle PEG coating including assessment of nanoparticle stability, particle uptake into neutrophils, neutrophil activation, and formation of reactive oxygen species as a result of nanoparticle exposure.

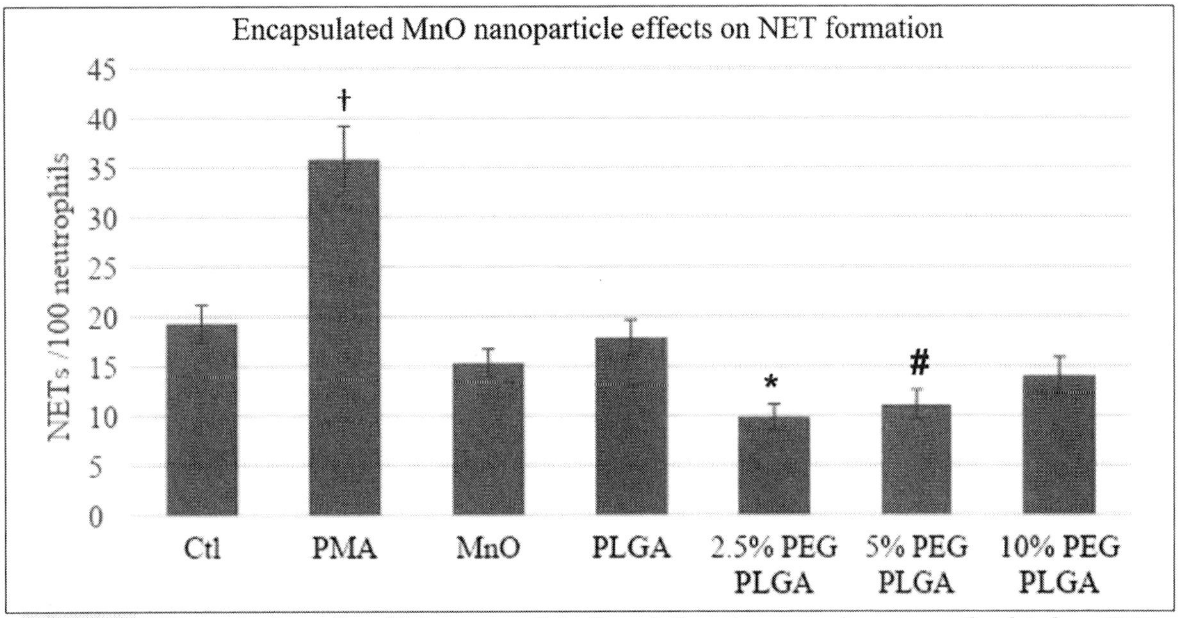

Figure 1: NET production of multiple nanoparticle formulations in comparison to unstimulated or PMA stimulated neutrophils. Error bars are standard errors. * indicates significance ($p<0.01$) relative to control. # indicates significance ($p<0.05$) relative to control. † indicates significance ($p<0.01$) relative to all other groups. Holm method was used to calculate p-values.

637764 Application of CFD-DEM in the Process and Pharmaceutical Industries

Abstract:

In this workshop, you would learn about Rocky DEM and the expanding range of granular systems that can be simulated accurately and effectively. Combined with CFD (Computational Fluid Mechanics), many processes across multiple industries can be simulated while representing real particle shape, particle size distribution and an accurate account of particle dynamic.
During the workshop presentation, industrial example and case studies will be used to highlight applications areas including:

- Slurry mills
- Cyclones, desanders, and drill cutting removal
- Pneumatic conveyors, hydraulic conveying
- Grain drying and sorting
- Tablet and candy coating
- Biomass reactors
- Fluidized beds and catalytic reactors

631442 Discrete Element Method Analysis for the Mechanochemical Grinding of Polymers

Abstract:

Plastic and synthetic polymer usage has increased at an accelerated rate since their mass production started in the 1950s. Since then, a cumulative volume of 6.3 billion tons (in 2015) has been produced and only 9% of them have been recycled [1]. Plastic monomers from fossil hydrocarbons generally cannot be consumed by microorganisms, which leads to an increased accumulation of plastics in landfills and oceans [2]. Thus, identifying new technical pathways that can efficiently process and recycle plastic waste is one of the greatest challenges of the next years for preserving the environment [1]. One pathway is mechanochemical depolymerization wherein the polymer chain is deconstructed by the application of some form of mechanical energy which then activates the chemical reaction. Traditionally, these reactions are performed in ball mills—tumbling mills where the grinding media are balls—which are known to have high energy consumption and hence are very expensive to operate. Therefore, accurate models, that describe the particles' interactions, based on physics, mechanics and chemistry are necessary for the optimization of their operation and reduction of expenses [3].

One promising route is Discrete Element Modeling (DEM). While there is rich history of research investigating DEM simulations for grinding applications such as in pharmaceuticals [4] and mining [5], such studies for depolymerizing plastics are as of yet very sparse. This submission adapts commonly used calibration tests to reproduce the kinematic interactions of a polymer ball-milling system via DEM. The resulting DEM simulation is exploited to identify phenomena most critical to and mechanisms uniquely apart of the control and optimization of polymer mechanocatalysis. Further, various reduced order modeling techniques are employed to develop faster and accurate representations of the DEM data, suitable for integration with population balance models for flowsheet modeling and optimization. Reduced-order models have played a significant role in connecting computational expensive models with optimization algorithms [6]. The techniques that will be presented employ Machine Learning regression and dimensionality reduction methods. In sum, this presentation aims to lay the foundation for DEM analysis and reduced-order modeling of polymer mechanocatalysis using ball-milling devices.

References

1. Štrukil, V., *Highly Efficient Solid-State Hydrolysis of Waste Polyethylene Terephthalate by Mechanochemical Milling and Vapor-Assisted Aging.* ChemSusChem, 2021. **14**(1): p. 330-338.
2. Geyer, R., J.R. Jambeck, and K.L. Law, *Production, use, and fate of all plastics ever made.* Science Advances, 2017. **3**(7): p. e1700782.
3. Wang, M.H., R.Y. Yang, and A.B. Yu, *DEM investigation of energy distribution and particle breakage in tumbling ball mills.* Powder Technology, 2012. **223**: p. 83-91.
4. Sen, M., et al., *A Multi-Scale Hybrid CFD-DEM-PBM Description of a Fluid-Bed Granulation Process.* Processes, 2014. **2**(1): p. 89-111.
5. Tavares, L.M., *A Review of Advanced Ball Mill Modelling.* KONA Powder and Particle Journal, 2017. **34**: p. 106-124.
6. Boukouvala, F., et al., *Reduced-order discrete element method modeling.* Chemical Engineering Science, 2013. **95**: p. 12-26.

629180 Effect of Aspect Ratio on Breakage Kinetics of Urea Crystals in Agitated Slurries

Abstract:

Previous research on paracetamol crystals investigated the effect on the particle size and shape distributions when the crystals were agitated in a nonsolvent. The initial parent crystals had aspect ratios near 2 in different size ranges. Experiments showed that the average aspect ratio in a given size range changed little as the residence time in the agitated was increased. Similar experiments were later performed with urea crystals that had higher aspect ratios. The urea crystals experiments in an agitated vessel also showed that the agitation residence time had little effect on the average aspect ratio of crystals in a given size interval.

While the urea crystal with the had a higher breakage rate than the paracetamol crystals, it is not clear that this was entirely due to the larger aspect ratio because the properties of the two substances are different. Additional urea crystal experiments were recently performed to provide more detailed information on the urea breakage kinetics. Specifically, the initial aspect ratios of the urea crystals used in the breakage experiments were narrower and more aspect ratios were used to aid in determining the effect of the aspect ratio on crystal breakage. These experimental results and the breakage distribution functions and parameters derived from the experiments are presented.

627007 Simulation of Full-Scale Open-Circuit, Multi-Compartment Cement Ball Mills: A New True Unsteady-State Simulator

Abstract:

The global demand for cement, as a common structural material for infrastructure, has progressively increased over the past decades, with world consumption already reaching 3.8 billion metric tons per annum by the early 2010s [1]. Cement manufacturing, however, involves energy-intensive processes that consume overall energy levels in the order of 100 kilowatt hours per metric ton of cement product. Of this consumption, 40% is accounted for by cement milling processes [2,3], where balling milling is used as the conventional cement milling technology. Ball milling can be performed under batch, continuous open-circuit, and continuous closed-circuit modes. Whereas batch ball milling is typically performed in a laboratory, continuous ball milling is operated as a full-scale process to achieve a high production rate [4]. In continuous open-circuit ball milling, feed material is continuously fed into the inlet of a mill and it is ground by grinding balls (see Fig. 1). Ball motion is initiated by the rotation of the shell, and the axial transport of the cement clinker particles is carried by the air stream and dispersion caused by ball–particle collisions. Balls hitting the powder bed results in breakage of the particles. Diaphragms (perforated partitions) are placed within and at the discharge end of the mill. As a rule of thumb [5], an open-circuit cement mill consists of three milling compartments with one discharge and two intermediate diaphragms. Particles finer than a diaphragm slot can pass onto the next compartment, whereas coarse particles are ground further until they fit into the slots. Intermediate diaphragms enable different degrees of milling (coarse and fine milling), whereas a discharge diaphragm mainly prevents grinding balls from being left from a mill. An internal mill shell is attached with a lining plate not only to prevent possible damage from grinding balls but also to increase milling performance. An example of this mechanism is a classifying liner, which appropriately segregates ball sizes along the length of an axial mill.

From the perspective of process engineering, the optimization of large-scale ball milling production by trial and error is costly and time-consuming. Without a strong fundamental understanding of this process, there is no guarantee that the best operation condition can be found. A reliable computer simulation would enable engineers to gain insight into complex milling processes and subsequently identify optimum operational conditions and bottlenecks in a system. Accordingly, in this study, we developed a novel true unsteady-state simulator (TUSSIM) on the basis of the solutions of a cell-based population balance model (PBM) (see Fig. 2). The cell-based discretization of the axial domain, along with the unknown cell residence time, leads to a set of differential-algebraic equations (DAEs). TUSSIM, coded in Matlab, provides an accurate, stable, and efficient numerical solution of the DAEs during both unsteady- and steady-state operations. Unlike other simulators [6,7], TUSSIM accounts for finite mixing, different ball mixtures, different ball size distributions (BSDs), internal classification in each compartment of ball mills, and the dynamic behaviors of the system.

In this study, TUSSIM was used to perform comprehensive simulations and explore the impact of ball sizes, ball mixtures, and number of compartments on the product size, which was reported in terms of specific surface area, median size, 10% passing size, and 90% passing size. First, the impacts of mixtures of different ball sizes (i.e., quaternary, ternary, and binary ball size mixtures as well as variable numbers of ball sizes) on product fineness were examined in comparison with the operations of a single ball size. Second, the quaternary ball size mixture was selected to investigate the impacts of three ball size distributions (BSDs): a uniform mass of balls (UM), a uniform number of balls (UN), and a uniform surface area of balls (USA). Moreover, the operation of the classifying liner in the last compartment of the mill was simulated to examine the benefits in terms of breakage kinetics. Finally, to assess why industry milling typically involves selecting three compartments for an open-circuit system, all simulations, except that on the BSDs, were repeated using a two-compartment configuration in which the second intermediate diaphragm of the mill was removed.

Overall, the results from TUSSIM simulations largely agree with various experimental observations and justify several common industry best practices. One of the important insights gained from them is that ball mixtures led to finer products than single ball size, consistent with industry practice [5]. Additionally, the binary mixture produced slightly finer products than those obtained using the ternary and quaternary mixtures, which aligns with experimental observations in lab-scale ball mills [8]. This finding, however, contrasts with industry practice, as three to five ball sizes are normally used [6,9]. This difference can be rationalized when ball wear is considered; using two considerably different ball sizes in a mixture may lead to a higher wear rate, and smaller balls can be potentially broken down by the larger balls. Hence, a binary mixture with two widely different sizes would not be favorable due to increased wear, albeit being favorable from a breakage kinetics perspective. Regarding the impact of BSDs, the UM distribution resulted in finer products than those generated via UN and USA, which is reasonable because UM involved more fraction of small balls compared with those used in UN and USA. The use of the classifying liner in the last compartment enabled suitable spatial variations in ball size along the axial length, resulting in finer products than those produced under spatially uniform mixed ball sizes. Finally, having the same length, the three-compartment mill produced a finer product that the two-compartment mill, and it showed less sensitivity to ball size selection. Overall, these simulations in view of existing cement milling practices suggest that TUSSIM is a credible simulation tool that can be effectively used for process simulation-optimization of open-circuit continuous dry ball mills.

References

[1] S.A. Miller, A. Horvath, P.J. Monteiro, Readily implementable techniques can cut annual CO_2 emissions from the production of concrete by over 20%, Environ. Res. Lett. 11 (2016) 074029.

[2] V. Ghalandari, A. Iranmanesh, Energy and exergy analyses for a cement ball mill of a new generation cement plant and optimizing grinding process: a case study, Adv. Powder Technol. 31 (2020) 1796-1810.

[3] A. Jankovic, W. Valery, E. Davis, Cement grinding optimisation, Miner. Eng. 17 (2004) 1075-1081.

[4] R.P. King, Modeling and Simulation of Mineral Processing Systems, Butterworth-Heinemann, Oxford, 2001.

[5] W.H. Duda, Cement Data Book, Volume One: International Process Engineering in the Cement Industry, 3rd ed. Bauverlag GmbH, Wiesbaden und Berlin, 1985.

[6] L.G. Austin, P.T. Luckie, D. Wightman, Steady-state simulation of a cement-milling circuit, Int. J. Miner. Process. 2 (1975) 127–150.

[7] S. Schwarz, J.M. Richardson, Modeling and simulation of mineral processing circuits using JKSimMet and JKSimFloat, SME Annu. Meet., 2013.

[8] S. Cayirli, Influences of operating parameters on dry ball mill performance, Physicochem. Probl. Miner. Process. 54 (2018) 751-762.

[9] O. Genc, Optimization of an industrial scale open circuit three-compartment cement grinding ball mill with the aid of simulation, Int. J. Miner. Process. 154 (2016) 1-9.

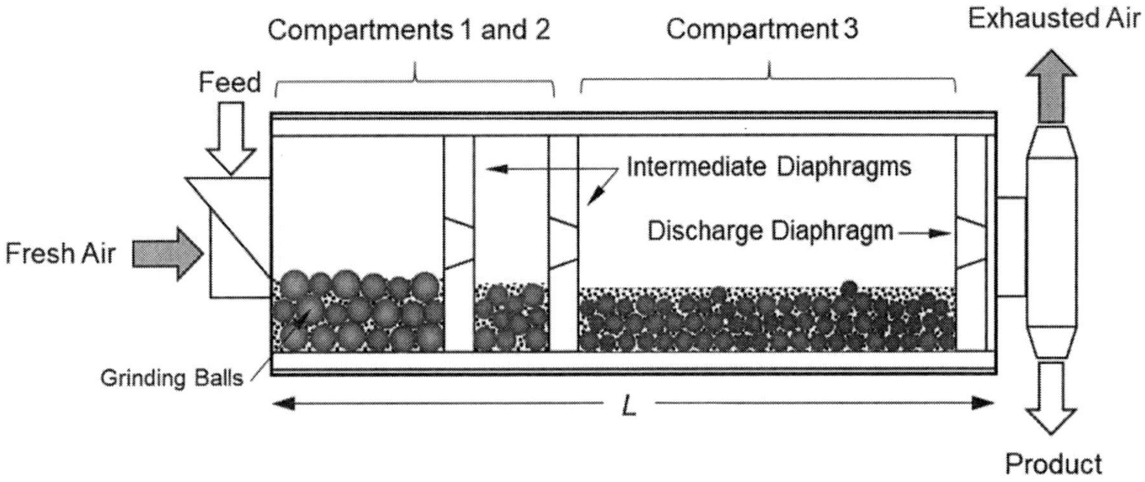

Fig. 1. A schematic of a 3-compartment open-circuit cement ball mill.

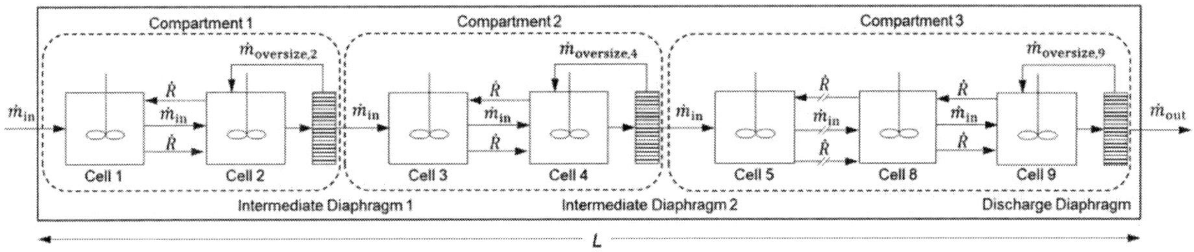

Fig. 2. A schematic of a 3-compartment open-circuit cement ball mill.

631305 Characterization of Polymeric Powders through Capillary Flow for Additive-Manufacturing Techniques

Abstract:

Binder jetting is an additive-manufacturing (AM) method that deposits a liquid binding agent onto a powder surface in a layer-by-layer fashion to form a solid part from a computer-aided-design (CAD) file. Powders that are utilized in AM include polymers, sands, and metals [1]. The production of parts requires input materials to have quantifiably repeatable characteristics and properties, which then requires standardized methods of characterization for powders [2]. In this investigation, we use a non-invasive frugal imaging technique to characterize fluid flow in a packed powder bed to investigate fluid-powder interactions at elevated temperatures. Laser-diffraction particle size and sieve analysis are used for size and shape analyses, as well as scanning electron microscopy (SEM) to investigate the surface topography. The results of this study demonstrate how particle size, size distribution, particle shape, surface roughness, and surface chemistry affect the packing density and therefore the binder-jetting process. The frugal technique that is presented in this study, which can be conducted only with a camera, an oven, and a commercially available microfluidic cell, offers a more affordable option to Washburn capillary rise method. As shown in figure, the trends from the preliminary experiments with the novel technique and the Washburn capillary rise method overlaps, which indicates that the novel technique is promising.

This study investigates the interactions of the polymer powders with fluids at the fundamental level. Results will aid in the understanding of other AM methods (e.g., cold spraying, drop on demand (DOD), selective laser sintering (SLS), electron beam melting (EBM)), which use polymer powders. Since this study uses a frugal technique, which requires only a mobile phone and a flow cell to analyze the fluid-powder interactions, characterization of the properties of any polymer powder is readily available to all industries [3].

References

[1] Prashanth Konda Gokuldoss, Sri Kolla, and Ju ̈rgen Eckert. Additive manufacturing processes: Selective laser melting, electron beam melting and binder jetting – selection guidelines. Materials, 10(6):672, 2017.

[2] Yun Bai, Grady Wagner, and Christopher B Williams. Effect of particle size distribution on powder packing and sintering in binder jetting additive manufacturing of metals. Journal of Manufacturing Science and Engineering, 139(8), 2017.

[3] Katrina J. Donovan. Microfluidic Investigations of Capillary Flow and Surface Phenomena in Porous Polymeric Media for 3D Printing. PhD thesis, Oregon State University, Corvallis OR, 2019.

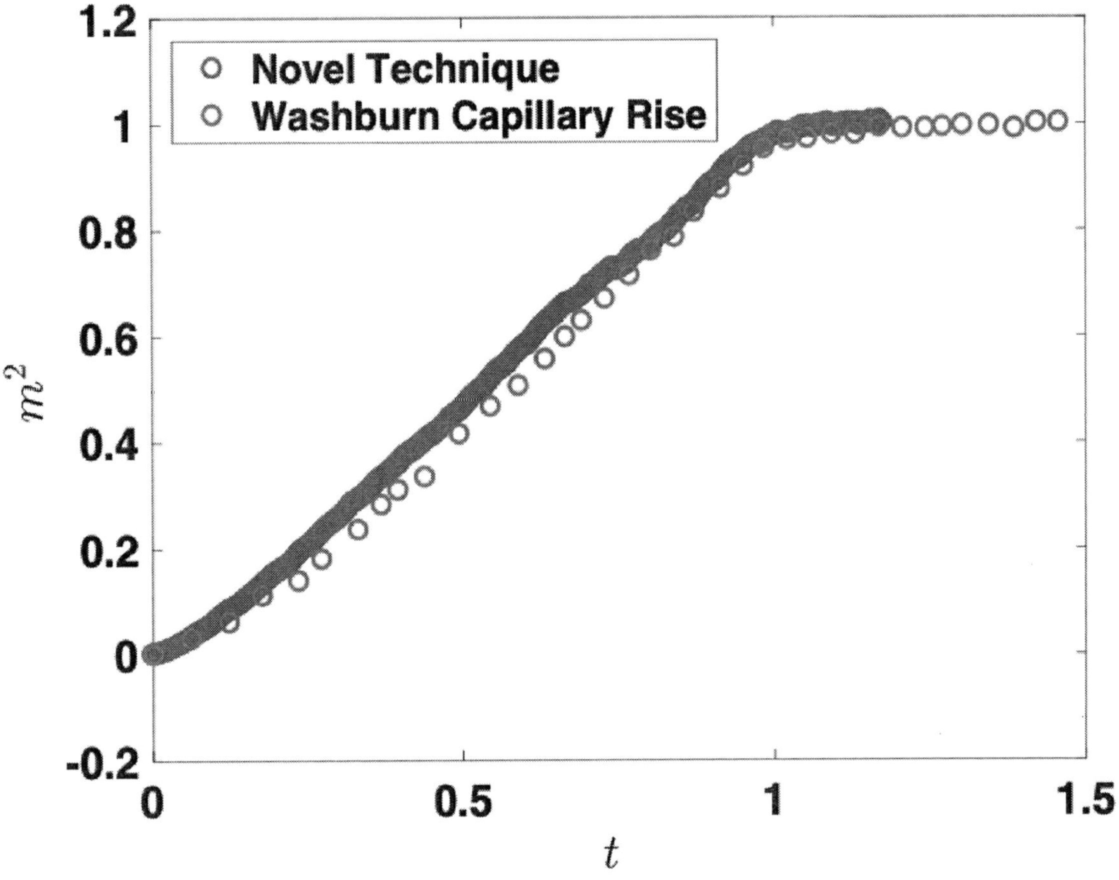

Figure. Results showing the overlap of Washburn capillary rise and novel method.

630167 Estimating the Three Characteristic Lengths of Plate-like Particles in Suspension

Abstract:

The size and shape of crystals are relevant properties of a powder, as they determine its rheology. In particular, ensuring a good filterability, flowability, and compressibility is crucial in the pharmaceutical industry, aiming for fast processing from crystallization to the final tablet manufacturing.

State-of-the-art characterization techniques, such as laser diffraction, focused beam reflectance measurement, or Coulter counter, typically rely on the assumption of sphericity, and therefore characterize the particles with a single size descriptor. Due to their lower symmetry, needle and plate-like crystals require two and three generic length descriptors, respectively, for an accurate size and shape characterization.

For the case of needle-like crystals, both mono and dual imaging systems have been successfully used to obtain the particle size and shape distribution[1,2]. Thanks to the larger amount of information gathered, dual imaging offers both improved accuracy and reliability. In addition, it is less sensitive to issues arising due to particle orientation with respect to the cameras. With the Dual Imaging System for Crystallization Observation (DISCO)[2], a flow-through device developed at ETH Zurich, we have been able to measure the particle size and shape distribution to a degree of accuracy such that the estimation of the suspended solid mass is made possible[3]. Further, the device has been successfully employed to estimate kinetics of fundamental phenomena and to effectively manipulate the size and shape of needle-like crystals[4,5].

For the case of plate-like crystals, to the best of our knowledge, no commercial device is able to accurately estimate their three characteristic lengths. This unavailability severely restricts designing processes to manipulate the size and shape of particulate populations with a plate-like morphology, even though they are frequently encountered in pharmaceutical processes. Recently, a neural network-based machine learning model built on the stereoscopic information provided by the DISCO has been proposed[6], thus laying the foundations for successful handling of particulate systems that require a 3D characterization.

In this contribution, we validate the aforementioned machine learning model and demonstrate that the DISCO can characterize plate-like particles with remarkable accuracy in terms of their three characteristic lengths, i.e., their length, width, and thickness.

Since no analytical standards are available for plate-like particles of different sizes and shapes, we produced populations of custom-designed "LithoPlatelets". The LithoPlatelets were fabricated by SU-8 photolithography. The SU-8 photoresist was spin-coated onto a silicon wafer to achieve the desired layer thickness, which determines the thickness of the platelets (in the order of tenths of micrometers). The structures were exposed through a photomask, and gently scraped off the wafer surface once fully polymerized[7]. Finally, to be measured with the DISCO, these reference platelet populations were suspended in mixtures of water and propan-2-ol. The measurement process is outlined in Figure 1.

Thanks to the flexibility of the fabrication method, several monodisperse population with a range of sizes and aspect ratios have been successfully characterized both individually and combined, thus proving the potential of the characterization technique when applied to real, polydisperse crystalline systems. We believe that this work therefore addresses two important challenges in the particle technology community. First, to exploit a reliable methodology to produce monodisperse populations of particles with nonstandard dimensions, which can be used as analytical references. Second, to provide a characterization platform featuring a robust, experimentally validated machine learning model that is able to accurately estimate the three characteristic lengths of plate-like particles.

[1] Cardona, J.; Ferreira, C.; McGinty, J.; Hamilton, A.; Agimelen, O. S.; Cleary, A.; Atkinson, R.; Michie,

C.; Marshall, S.; Chen, Y.-C.; Sefcik, J.; Andonovic, I.; Tachtatzis, C. Image Analysis Framework with Focus Evaluation for in Situ Characterisation of Particle Size and Shape Attributes. *Chem. Eng. Sci.* **2018**, *191*, 208–231.

[2] Rajagopalan, A. K.; Schneeberger, J.; Salvatori, F.; Bötschi, S.; Ochsenbein, D. R.; Oswald, M. R.; Pollefeys, M.; Mazzotti, M. A Comprehensive Shape Analysis Pipeline for Stereoscopic Measurements of Particulate Populations in Suspension. *Powder Technol.* **2017**, *321*, 479–493.

[3] Bötschi, S.; Rajagopalan, A. K.; Morari, M.; Mazzotti, M. An Alternative Approach to Estimate Solute Concentration: Exploiting the Information Embedded in the Solid Phase. *J. Phys. Chem. Lett.* **2018**, *9* (15), 4210–4214.

[4] Salvatori, F.; Binel, P.; Mazzotti, M. Efficient Assessment of Combined Crystallization, Milling, and Dissolution Cycles for Crystal Size and Shape Manipulation. *Chem. Eng. Sci. X* **2019**, *1*, 100004.

[5] Bötschi, S.; Rajagopalan, A. K.; Rombaut, I.; Morari, M.; Mazzotti, M. From Needle-like toward Equant Particles: A Controlled Crystal Shape Engineering Pathway. *Comput. Chem. Eng.* **2019**, *131*, 106581.

[6] Jaeggi, A.; Rajagopalan, A. K.; Morari, M.; Mazzotti, M. Characterizing Ensembles of Platelike Particles via Machine Learning. *Ind. Eng. Chem. Res.* **2021**, *60* (1), 473–483.

[7] Wyatt Shields Iv, C.; Zhu, S.; Yang, Y.; Bharti, B.; Liu, J.; Yellen, B. B.; Velev, O. D.; López, G. P. Field-Directed Assembly of Patchy Anisotropic Microparticles with Defined Shape. *Soft Matter* **2013**, *9* (38), 9219–9229.

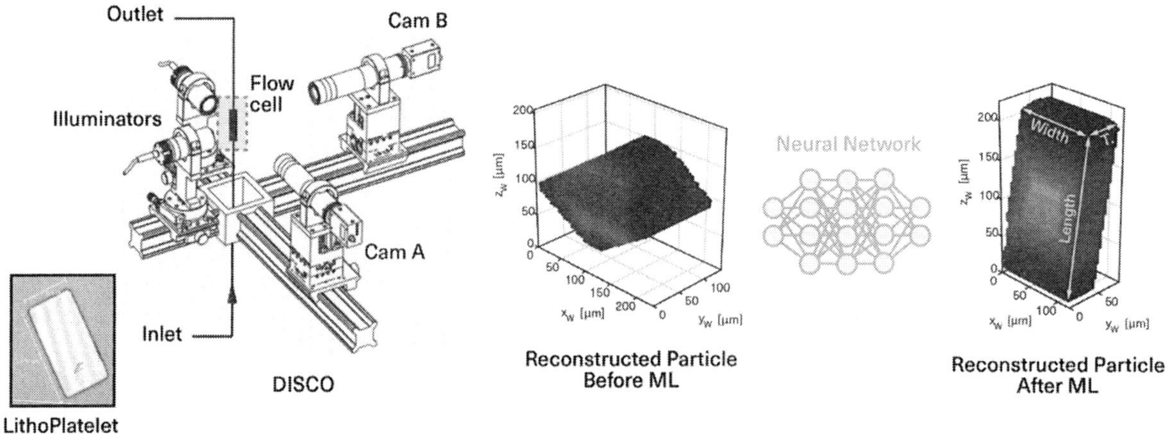

Figure 1. Platelets made of SU-8 through a photolithographic process have been used as analytical standards. The suspension flows through a quartz flow cell and images are collected with two cameras arranged 90° apart from each other. The pairs of images are used to build the visual hull of the imaged objects, which for plate-like particles may not match their true 3D shape. The metrics extracted from the visual hull and from the single images are processed with a neural network-based machine learning (ML) model, which had been previously trained on in-silico data, to yield more accurate estimates of the length, width, and thickness of the platelets.

629594 Predicting the Behavior of Different Tablet Press Feed Frame Systems Using the Discrete Element Method (DEM) Modeling

Abstract:

Tablets are the most common dosage form for pharmaceutical products. For bulk tablet manufacturing, rotary tablet presses are commonly used to compress powder blends containing active pharmaceutical ingredients (APIs). However, prior to the compression step the powder blend must be fed from a hopper to a feed-frame system before it reaches the die table of the tablet press. Consideration is required when transferring a compression process, such as encountered when moving to a different press of the same scale or during scale-up. Tablet presses vary in their feed-frame design which can result in different powder flow dynamics affecting the final drug product. In order to efficiently transfer the process from one equipment to another it is important to understand performance differences in the behavior of different tablet press feed frame systems. In this study, the discrete element method (DEM) is used to study the differences in powder flow dynamics across different rotary tablet presses. The effect of paddle speed, die table speed, particle and wall properties on the resulting powder flow behavior, feed-frame residence time, and tablet weight variability are investigated.

627735 Fundamental Investigation of the Rheology and Structure of Capillary Suspensions Made with Oyster Particles

Abstract:

Ternary particle-liquid-liquid (PL2) suspensions offer different properties from their elementary constituents or components. One quadrant of these systems are the particle-stabilized emulsions also known as Pickering emulsions [1, 2]. Pickering emulsions offer unique features and have found great potential in agrochemical applications, as well as in the oil, food and paint industries while pharmaceutical and cosmetic products are increasingly explored [3,4]. The formation of a stable emulsion requires a mechanically balanced and sufficiently resistant film that will occlude the dispersed phase from the continuous. This is mostly established by the alignment of amphiphilic surfactant molecules on the droplet's interface. Similarly, particles tend to position themselves on the interface based on their hydrophilic or hydrophobic surface, their wettability profile, size and shape.

The depth of immersion of a particle in the liquid depends on the angle of contact shown between the liquid-particle-liquid interface [5] while the particles' morphology plays also a role [4]. Previous studies have shown that the tendency to destabilize depends strongly on the degree of surface coverage, which with increasing particle concentration can transition from low surface coverage, to monolayers and multilayers and reach up to three dimensional network structures [6].

In the present investigation we review the evolution of Pickering emulsions and focus on the innovative use of oyster shell particles for the stabilization as our base case study. According to FAO* forecasts the consumption of oysters is expected to rise and with that the untreated volume of shells will also increase. By incorporating oyster shell particles in our formulations, we present a potential solution to the waste accumulation by considering the increased demand for clean label products and emulsifier-free formulations and at the same time we suggest these systems as potential edible particle stabilizers in the food sector.

*Food and Agriculture Organisation

Wet grinded oyster shells were prepared and tested in a cream formulation at different concentrations. The evaluation period lasted eight weeks and the Pickering emulsions were analysed by particle size measurements, instability indexes and characterised using microscopy. Based on the results the formulations containing concentrations ranging from 6 wt-% to 8 wt-% oyster shell particles showed great potential. The formulation showing better performance was then scaled up and compared with other formulations including aerosil and other surfactants and reviewed for a further period of eight weeks. The analytical part was extended, and rheological measurements rounded up our understanding of their stability. Among these, viscosity curves, the yield point and extrapolated long-term storage behaviour were analysed by means of rotation and oscillation measurements (Physica UDS 200; Anton Paar).

Figure 1a depicts the preparation process followed for the Pickering emulsion formulation by using an Ultra-Turrax ®.

Full details of the methods employed, the control experiments and the results obtained will be presented.

Figure 1b shows the results obtained for the viscosity dependence with respect to shear rate for time variations considering the different formulations. These profiles show a decrease in dynamic viscosity with increasing shear rate and the curves are characteristic of shear thinning behaviour. This is to be expected with emulsions because the formation of finely distributed oil droplets occurs and thus the internal friction decreases. The fact that the dynamic viscosity does not change much over time (for eight weeks) indicates that the structure of the emulsions and the oil droplets remain stable.

Figure 1b. Viscosity dependence with respect to shear rate for time variations considering the different formulations.

This research bridges the knowledge gap on the structural transitions of ternary suspensions. Furthermore we also point out opportunities in the current research on the applications of Pickering emulsions as capillary suspensions and suggests future directions for the design of particulate stabilizers and preparation methods for Pickering emulsions and their derived materials. [7]

References

[1] Ramsden W. Separation of solids in the surface-layers of solutions and 'suspensions' (observations on surface-membranes, bubbles, emulsions, and mechanical coagulation).—Preliminary account 1903:156–64.

[2] Pickering SU. Emulsions 1907:2001–21.

[3] Aveyard R, Binks BP, Clint JH. Emulsions stabilised solely by colloidal particles. Advances in Colloid and Interface Science 2003;100-102:503–46.

[4] Huang F, Liang Y, He Y. On the Pickering emulsions stabilized by calcium carbonate particles with various morphologies. Colloids and Surfaces A: Physicochemical and Engineering Aspects 2019;580:123722.

[5] Inoue M, Hashizaki K, Taguchi H, Saito Y. Formation and characterization of emulsions using beta-cyclodextrin as an emulsifier. Chem Pharm Bull (Tokyo) 2008;56(5):668–71.

[6] Schröder A, Sprakel J, Schroën K, Spaen JN, Berton-Carabin CC. Coalescence stability of Pickering emulsions produced with lipid particles: A microfluidic study. Journal of Food Engineering 2018;234:63–72.

[7] Jiang H, Sheng Y, Ngai T. Pickering emulsions: Versatility of colloidal particles and recent applications. Curr Opin Colloid Interface Sci 2020;49:1–15.

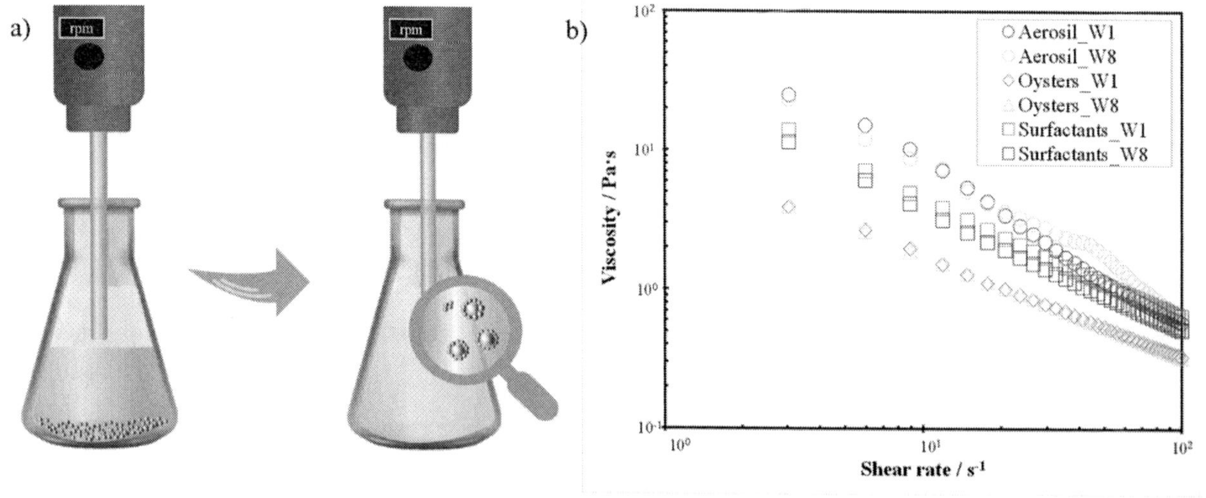

637762 Novel Altair Simulation Solutions for Particulate and Particle Laden Flows

Abstract:

Altair EDEM is a discrete element method solver which enables users to accurately simulate particulate processes. Coupled to Altair Computational Fluid Dynamics (AcuSolve) this also enables the simulation and analysis of two-phase solid-fluid systems.

These simulation tools are widely used in various industries and academia to deepen the understanding of different unit operations such as mixers, coaters, fluidized beds and pneumatic conveyors. EDEM and coupled AcuSolve-EDEM provide particle-level information about processes that can be impossible or extremely challenging to obtain through experimental studies. Despite the great potential of these tools for investigating various processes, there are some shortcomings when employing them.

In this workshop, a brief overview of EDEM and coupled AcuSolve-EDEM is presented. A few industrial test cases performed by EDEM and AcuSolve-EDEM are demonstrated. Some shortcomings associated with these tools and DEM and CFD-DEM simulation techniques in general are discussed. These challenges include long computational time, the size and number of particles which can be simulated in a reasonable computational time, particle shape and the material calibration process. The unique and novel solutions that Altair offers in order to address these challenges when using DEM and coupled CFD-DEM in various industries are presented.

632580 Spray Drying of Drug Loaded Nanoparticles with Matrix Forming Excipients

Abstract:

Nanoparticle drug delivery systems are of recent increased interest for applications in both oral and parenteral drug delivery. Typically, nanoparticles are formed in liquid media by precipitation or self-assembly, resulting in a nanoparticle suspension. However, it is often desirable to obtain a dry powder form of the nanoparticles that may be later resuspended in liquid media. The dry powder facilitates efficient storage and transport of the nanoparticle formulation.

One high-throughput method of achieving a dry powder from a liquid nanoparticle suspension is via spray drying. During spray drying, a heated drying gas contacts atomized liquid droplets, promoting fast evaporation of the liquid and resulting in a dry powder. Spray drying is commonly employed in many large-scale processes in the food processing and pharmaceutical industries, notably for the production of dry milk powder. However, in the case of nanoparticle therapeutic formulations, the addition of matrix-forming excipients during spray drying is necessary in order to preserve the integrity of the nanoparticles during the drying process. These excipients result in the formation of micron scale matrix particles containing embedded nanoparticles. When reconstituted in a liquid, the matrix dissolves, thus re-dispersing the nanoparticles.

Here, we present a systematic approach to investigate the use of various matrix materials during the spray drying process, including functionalized cellulose, sugars, and synthetic polymers. We study the effect of matrix former on the size and polydispersity of the spray-dried nanoparticles upon re-dispersion. By varying the mass ratio of matrix former to nanoparticles, we study the amount of matrix former required to prevent nanoparticle fusion and aggregation during the drying process. We employ model nanoparticles to investigate the effect of the nanoparticles on the thermal properties of the resulting matrix particles. Polymer-stabilized core-shell nanoparticles with core materials of varying glass transition temperature are spray dried, and the matrix particles are characterized by calorimetry. We show that the composition of the nanoparticles influences the thermal properties of the matrix particles.

632747 Methods for Encapsulating Mobile Microparticles

Abstract:

Particle containing composite microcapsules have become increasingly prevalent and are widely employed in a wide range of applications including pharmaceutical, personal care, nanocomposite self-healing barrier films and coatings industries. Most commonly, small particles are held at the interface (as seen in colloidosomes), or physically arrested in a solid core. Microencapsulation through polymerization or precipitation on the interface of emulsion droplets, has been widely successful in creating capsules of fluids that can be then incorporated in other media. However, due to surface energy, Pickering stabilization, or adsorption of particles at interfaces, it is difficult to synthesize microcapsules of suspensions of freely diffusive or movable particles. Various examples exist, however they typically lack robustness in their material and physical properties.

We present various approaches to encapsulating spatially manipulatable microparticles, over a broad particle size range (1-50 microns in diameter), within the core of a microcapsule, avoiding substantial Pickering stabilization and adsorption. One approach uses an emulsified yield stress suspensions encapsulated through a standard polyurea interfacial polymerization. This yield stress material is tuned to match the physical properties of the encapsulated microparticles to hinder particle diffusivity to the capsule walls during polymerization and to balance gravity induced settling, whilst also not permanently fixing particle locations. Particles are actively released through either a temperature change or by manipulation of particles through external forces. This was demonstrated with a magnetic field for Janus particles functionalized with a ferromagnetic coating of iron oxide to directly visualize particle translation and rotation. A second approach looks at the use of density matching between binary solvents and microparticles in polyurea based microencapsulations to achieve a similar effect without the need for an external activating force. Both methods prove viable over a range of particle sizes and could be applicable to a number of composite microcapsule applications.

This work was supported by the Applied Physics Laboratory at Johns Hopkins University.

622442 Magnetic Nanomaterial Endocytosis: Uptake and Cytotoxicity

Abstract:

Magnetic nanomaterials have found recent popularity in endocytosis, cell separation, drug delivery, tumor targeting, biomedical imaging, and in vivo diagnostics. Endocytosis (internalization by cells) of magnetic nanoparticles is of fundamental importance to a broad range of biomedical applications depending on the nanoparticle size, shape, surface functionalization, and magnetic properties. Magnetic cytometry captures the motion of labeled cells in an isodynamic magnetic field to estimate the magnetophoretic mobility of labeled cells. Magnetophoretic mobility is proportional to the number of beads ingested per cell. Using magnetic cytometry, it is possible to estimate the number of particles ingested by each cell within populations of thousands of cells. We employed a Hyperflux^{TM} particle tracking velocimeter to measure magnetophoretic mobility of CHO cells in monolayer culture ingesting 50 nm and 100 nm magnetic beads (fluidMAG^{TM}, Chemicell) with different surface coatings. In addition to uptake data, cell loss and viability data are available in the course of endocytosis studies by this method. Reduction in cell number is correlated with the level of cellular uptake. Positively charged beads resulted in the highest level of uptake and the greatest reduction in cell number, while PEG-coated beads were the least ingested and the least toxic. Maximum cell losses observed were 50% on day 1 and 65% on day 2 of incubation at the highest concentration of beads studied corresponding to 200 µg of iron/mL. The viability of cells that survived processing was evaluated by viability staining methods, including flow cytometry. In most cases, >80% of cells were viable after one day of exposure to 100 µg of iron/mL by staining criteria. The results in general follow a previously found rule that cell loss is proportional to magnetic nanomaterial uptake, and this principle can guide the use of nanomaterials in biomedical applications.

630170 Engineering Nanostraws for MCM-41 Microparticles Towards Enhancing the Adsorption of Guest Molecules: Application in CO_2 Capture.

Abstract:

Ordered mesoporous silica particles such as MCM-41 have wide use in molecular adsorption and reaction catalysis due to their well-defined structure, porosity, and high surface areas. Spherical particles of MCM-41 are easily synthesized using a one-step facile aerosol assisted technique. However, the tight 3 nm pore channels of MCM-41 pose a challenge to molecules penetrating the interior of the pellet and this diminishes the efficiency of this material in adsorption and catalytic applications. We report new finding on introducing nanostraws into MCM-41 pellets to mitigate this diffusional restriction. We demonstrate the use of the aerosol synthesis to systematically incorporate clay nanotubes known as halloysites into MCM-41 pellets in one step. The morphology of the composite material is such that the nanotubes extend from the interior and protrudes through the surface of the pellet like straws. These ceramic nanotubes are 0.5 - 3 μm long and their lumen diameter ranges from 15 - 40 nm which are much larger than the pores of MCM-41. Thus, the nanotube lumen should provide a larger pathway for molecules to access the interior of the MCM-41 pellets. Halloysite is an environmentally benign material that is abundant and inexpensive, therefore it does not add significant costs to the production of this composite material. The composite MCM-41 particles with nanostraws are highly reproducible and detailed characterization showed that the intrinsic structural properties of the MCM-41 remain unchanged with the incorporation of the nanostraws. To illustrate the concept of improved molecular transport into MCM-41 pellets enabled by the nanostraws, the composite material was studied as a solid adsorbent for the impregnation of polymeric amines for CO_2 capture. In comparison to MCM-41, the composite with the nanostraws showed improved loading of polymeric amine macromolecules and consequently doubled the CO_2 capture capacity of the impregnated amines.

630550 Characterizing Graphene Oxide Suspensions with Rheological Testing and Modeling, Neutron Scattering, and Electrochemical Performance

Abstract:

Electron transfer in carbon based colloidal particle systems make them a good candidate for a variety of electronics applications from flowable batteries to conductive coatings. In order to develop a toolbox to design application driven conductive suspensions, it is important to be able to understand primary particle morphology, surface chemistry, aggregate morphology and network structure. To this end, our group has synthesized graphene oxide using the modified Hummer's method[1] to use as the basis for our study in colloidal particles in homogenous suspension after ultracentrifugation. The surface chemistry of the graphene oxide in aqueous solution determines the degree of aggregation and the formation of hierarchical structures. The inherent graphene properties and concentration (loading) affect the electronic conductivity, mass transport, and stability of the microstructure in flowable electrochemical applications. Therefore, we have characterized the graphene oxide using electrochemical techniques such as cyclic voltammetry and impedance spectroscopy to evaluate its electronic and ionic conductivity.

Moreover, to understand the evolving microstructure, and its contribution to conductivity we incorporated rheological testing. The evolving microstructure of the graphene contributes to particle (sheet to sheet) interaction. Rheological data shows shear thinning with the graphene suspensions In addition, we show both the linear Small Amplitude Oscillatory Shear and nonlinear Large Amplitude Oscillatory Shear flow experiments as well as results of fitting and predicting the rheological flows with contemporary thixo-elasto-visco-plastic (TEVP) models. We will present our ultra-Small Angle Neutron Scattering results with USANS. From this we hypothesize that our material structure is linked to a breakdown of a hierarchical fractal structure in these suspensions. Preliminary Small Angle Neutron Scattering (SANS) and USANS measurements seem to validate this hypothesis. We show the graphene oxide suspension has a mid q feature that can be associated with the primary particle structure and a large upturn at low-q (inverse aggregate length scale) due to the formation of larger structures in the sample. In order to fully characterize the graphene-oxide we show a combination of SANS and USANS. We plan extract primary particle size, aggregate size and fractal dimension from the SANS/USANS data and correlate these properties directly to the conductivity and linear rheological properties.

Lastly, we will demonstrate a correlation between our rheological model's prediction of thixotropy, or microstructure with elastic and viscous features of our transient rheological data using Sequence of Physical Processes (SPP). We will use SPP and the TEVP modeling to show mechanical differences in the preparation (500g vs. 1000g formulations). We will show how the evolving microstructure contributes to the mechanical properties of the material with Cole-Cole plots. In addition, we will construct the full graphene oxide characterization with the USANS/SANS, rheological modeling, SPP framework analysis, and conductivity of the microstructure itself.

References

[1] Yu et al. Sci Rep (2016)

[2] Horner et al. JOR (2019).

[3] Rogers Rheo. Acta (2017).

[4] Richards et al. JOR (2017).

[5] Hipp et al. JOR (2019).

631485 Binder-Free Twin-Screw Melt Granulation: An Effective Approach to Manufacture High-Dose API Formulations

Abstract:

Purpose: Granulation is an important process step applied in manufacturing of numerous pharmaceutical products because it improves flowability, compressibility, content uniformity and, in some cases, the dissolution rate of the materials once they are formulated into a final solid dosage form (e.g., tablets). Granulation processes are usually classified according to the binder nature as wet, dry or melt. The absence of water or other solvents in melt granulation processes makes this approach quite valuable for moisture-sensitive ingredients and since it completely avoids the disadvantages associated with solvent recovery and final disposal. Among the available granulation technologies, twin-screw granulators have recently brought attention to both academic scientists as well industrial researchers since they present a homogeneous powder and energy distribution within the unit to every portion of material processed, resulting in a consistent product quality even at low binder concentration levels. Furthermore, as an inherently continuous process, twin-screw granulators can be readily integrated into the continuous manufacturing of pharmaceutical dosage forms using existing continuous granulation equipment. In this context, this study investigates the use of twin-screw binder-free melt granulation (BFMG) in the development of high-dose solid dose formulations for low melting point thermally stable drugs. Ibuprofen and guaifenesin, which are commonly used in high-dose formulations, were chosen as model drugs to test this novel application of twin-screw melt granulation. **Method:** A design of experiments (DoE) response surface methodology was used to establish the design space for the end-product. The effects of the most relevant process variables (barrel operating temperature, powder feed rate, screw speed and screw configuration) on granule properties (outlet temperature, size distribution, morphology, flowability, compressibility, porosity) and tablet (600 mg) attributes (tensile strength and in-vitro dissolution) were thoroughly studied. **Results & Conclusion:** By granulating pure API powder, it is shown that BFMG can successfully be used to produce granules that contain 100% API. Furthermore, barrel temperature (alone or in interactions with the other variables) represented the most significant variable for both drugs since it governs the formation of granules by partial melting and subsequent agglomeration of the fed powder. Interestingly, the shear action originated by screw speed and screw configuration resulted in various significant responses depending on the drug substance, indicating that it can also be affected by the nature of the processed molecule. Flow properties were improved (i.e., lower Hausner ratio) for both drugs after granules formations. Tabletability was also tested by preparing tablets for all samples. Surprisingly, the resulting granules were highly compactible, requiring only 1% lubricant to form strong tablets containing 99% API. The results also showed that tablets become harder as the granule size increased, especially for guaifenesin. Different increasing behavior was observed for both drugs, indicating that a thorough analysis should be performed regarding the nature of the generated bonds between particles and the granules strength. Based on previous experience, this effect should level off for a certain granule size due to the likely formation of failure-inducing voids for very large granules. As expected, in-vitro dissolution results indicated that tablets and capsules showed slightly slower dissolution rates than the granules. Nevertheless, all dissolution profiles were in agreement with the USP requirements for immediate release dosage forms, being faster the release of guaifenesin since its water solubility is higher than for ibuprofen. In conclusion, it is proved that BFMG of low melting point drugs can be used in to make tablets that are almost 100% API with excellent mechanical and dissolution behavior. This suggests that BFMG can be an important enabling technology for many high dose pharmaceutical products.

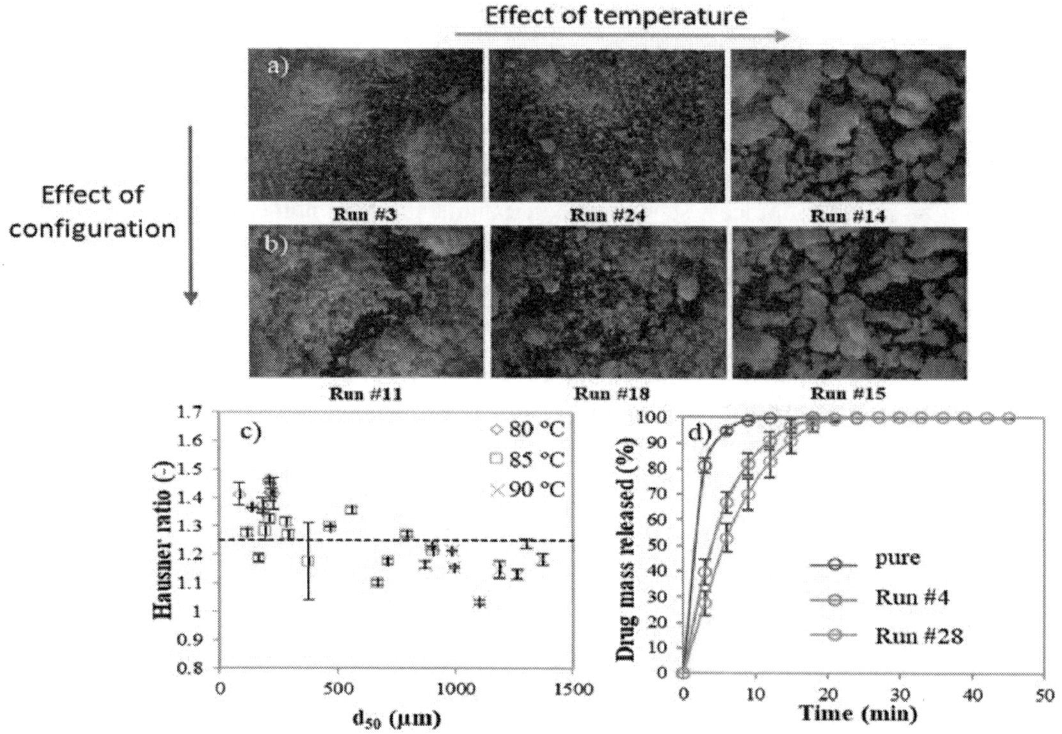

Figure 1 (a) Selected runs granules for guaifenesin obtained with 4 kneading zones varying operating temperature, (b) 6 kneading zones varying operating temperature. (c) Hausner ratio versus mean size, and (d) dissolution rate studies of tablets made of pure powder or granules.

631874 A Mechanistic Analysis on the Effect of Mixing Dynamics on Granule Microstructure

Abstract:

Granule microstructure is an important critical quality attribute. It encumbers many individual properties like porosity, surface morphology, component content uniformity, pore size distribution and, open to closed pore ratio. All these properties combined not only affect the flow, strength, content uniformity, and stability of granules [1, 2] but also affect the critical quality attributes of downstream products such as tablets and capsules [3, 4].

Twin-screw granulation (TSG) is a widely used continuous wet granulation technique. In the twin-screw granulator, the granulation regimes are physically separated along the length of the screws [5]. Moreover, given the lack of space available for the granules to grow in a twin-screw granulator, the stresses per unit time applied on the granules are higher compared to high shear and fluid bed granulators [6]. As a result of these factors, the granule formation physics in a TSG are significantly different compared to other types of granulation techniques [5].

Since the physics involved in twin-screw granulation is different from other types of granulation techniques, the factors affecting the granulation rate mechanisms and thence the critical quality attributes of granules are quite different. Studies focusing on explaining the relationship between the input factor, granulation rate mechanism, and granule quality attribute are few, especially for granule microstructure.

Mixing dynamics control the distribution and interaction of particles inside the granulator, which in turn govern the granulation rate mechanisms and ultimately the critical quality attributes of the granule [7]. In an earlier study by the authors, a quantitative relationship between input process & design parameters and mixing dynamics has been established [8]. The objective of this study is to identify and study in detail the different relationships between mixing dynamics, granulation rate mechanisms, and granule microstructure in a twin-screw granulator. In particular, the focus will be on studying the component distribution within the granule, porosity of the granule and the pore network distribution within the granule. All these properties will be studied quantitatively.

In this study, the fundamental factors affecting the granule microstructure are hypothesized. Appropriate experimentation and analysis are underway, and the results of this analysis will be used to accept or reject the hypotheses. Four different hypotheses, linking the mixing dynamics to granulation rate mechanism and ultimately to granule microstructure, are proposed as part of this study.

Hypothesis 1 (Fill level): The powder fill level in the granulator controls the amount of powder in the granulator which affects the number and frequency of collisions of powder particles in the granulator. This collision frequency influences the strength of the shear and compaction forces on the powders and alters the granule microstructure. Granules become denser with an increase in shear forces inside the granulator until a critical force. Shear forces higher than the critical force result in the breakage of granules. The fill level in the granulator will be varied by carefully manipulating the screw speed, throughput, and screw design.

Hypothesis 2 (Liquid Bridges): Liquid bridges are formed when two saturated granules form a bigger granule through successful collision. The collision is deemed successful when the viscous dissipation force between the granules(due to the liquid bridge) is greater than the collision kinetic energy. Liquid bridge strength between granules affects the density and strength of the granule microstructure. These bridges are affected by the solid-liquid distribution and viscosity of the liquid binder. The effect of liquid bridges will be varied by varying the liquid to solid ratio and viscosity of the binder.

Hypothesis 3 (Particle/Agglomerate mixing): Particle/agglomerate mixing affects the component uniformity across different granule size fractions and also within a single granule. The type of colliding-granules also affects the density of the resultant larger granule being formed due to the shape and state of saturation of the colliding granules. The state of mixing also affects the growth mechanism. Particle/agglomerate mixing is a direct consequence of mixing dynamics and will be varied carefully by varying the screw speed, throughput, and screw design.

Hypothesis 4 (Powder Wettability): Mixing dynamics present in the granulator affect the type of particles present in the vicinity of the incoming droplet. This affects the nucleation mechanism, especially in a multi-component blend consisting of both hydrophilic and hydrophobic powders. It was observed in the literature that, hydrophilic material forms smaller, denser nuclei via immersion nucleation and hydrophobic nuclei form larger, hollow nuclei via the solid-spreading mechanism [9]. The kind of nuclei will in turn affect the granulation rate mechanism and the microstructure of the granule formed [7].

By quantitatively studying the component distribution and pore network distribution, this study aims to further the mechanistic understanding of the effect of various process, design, and material parameters on the granule microstructure. Moreover, this study will help in establishing the fundamental correlation between mixing dynamics, granulation rate mechanisms, and granule microstructure.

Figure 1: Map summarizing the effect of mixing dynamics on the granule microstructure.

1. Oka, S., et al., *The effects of improper mixing and preferential wetting of active and excipient ingredients on content uniformity in high shear wet granulation.* Powder Technology, 2015. **278**: p. 266-277.
2. Ansari, M.A. and F. Stepanek, *The effect of granule microstructure on dissolution rate.* Powder Technology, 2008. **181**(2): p. 104-114.
3. Kashani Rahimi, S., et al., *The role of the screw profile on granular structure and mixing efficiency of a high-dose hydrophobic drug formulation during twin screw wet granulation.* International Journal of Pharmaceutics, 2020. **575**: p. 118958.
4. Meng, W., et al., *Statistical analysis and comparison of a continuous high shear granulator with a twin screw granulator: Effect of process parameters on critical granule attributes and granulation mechanisms.* International Journal of Pharmaceutics, 2016. **513**(1): p. 357-375.
5. El Hagrasy, A.S. and J.D. Litster, *Granulation rate processes in the kneading elements of a twin screw granulator.* AIChE Journal, 2013. **59**(11): p. 4100-4115.
6. Dhenge, R.M., et al., *Twin screw wet granulation: Effects of properties of granulation liquid.* Powder Technology, 2012. **229**: p. 126-136.
7. Mundozah, A.L., et al., *Assessing Particle Segregation Using Near-Infrared Chemical Imaging in Twin Screw Granulation.* International Journal of Pharmaceutics, 2019. **568**: p. 118541.
8. Lalith Kotamarthy, R.R., *Mechanistic understanding of the effects of process and design parameters on the mixing dynamics in continuous twin-screw granulation.* Powder Technology, "*under review*".
9. Muthancheri, I. and R. Ramachandran, *Mechanistic understanding of granule growth behavior in bi-component wet granulation processes with wettability differentials.* Powder Technology, 2020. **367**: p. 841-859.

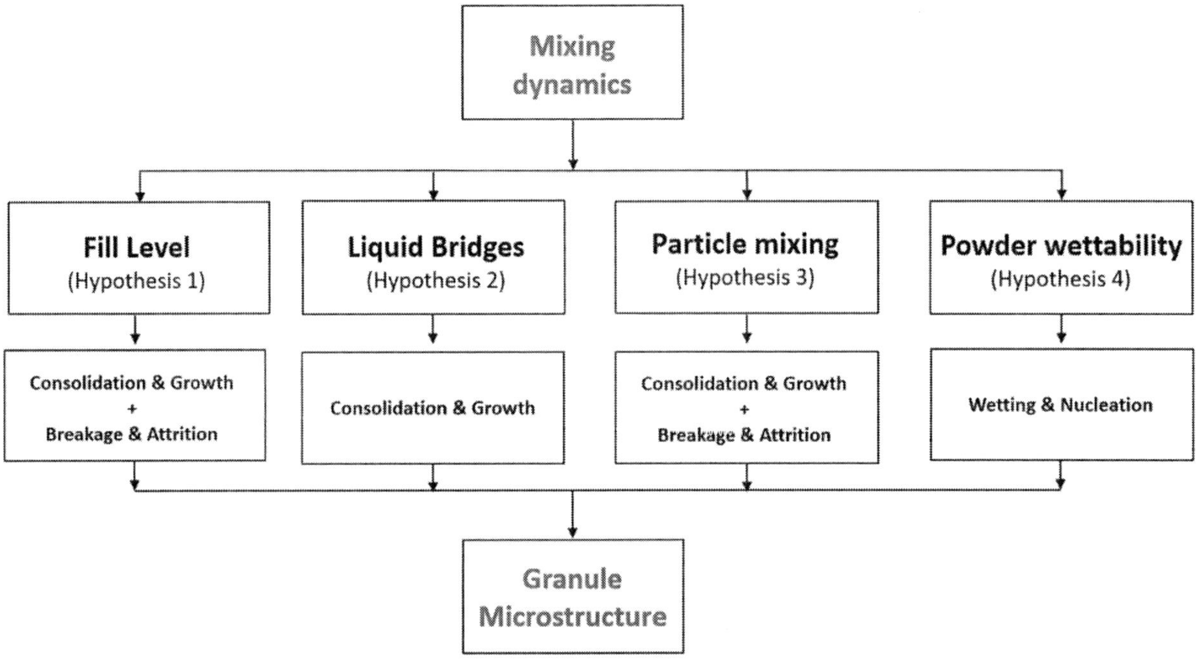
Mixing dynamics
Fill Level
(Hypothesis 1)
Liquid Bridges
(Hypothesis 2)
Particle mixing
(Hypothesis 3)
Powder wettability
(Hypothesis 4)
Consolidation & Growth
+
Breakage & Attrition
Consolidation & Growth
Consolidation & Growth
+
Breakage & Attrition
Wetting & Nucleation
Granule Microstructure

620422 Controlled Release of Urea Composites through Formulation and Process Design

Abstract:

The high solubility in water and porous structure of urea applied in crop fertilization result in a 50 % loss of its content through ammonia volatilization, a greenhouse gas, and nitrate leaching. Controlled release technology has the potential to mitigate the environmental and health impact due to urea fertilizer losses. This work aims to investigate an underexplored area: the mechanistic influence of formulation and process design on the pore internal structure and the release rate of urea composites. Urea and four binders in powder form (hydroxypropyl methylcellulose grade E5- HPMC, high molecular weight hydroxypropyl cellulose- HPC, gluten from wheat, and a mixture of xanthan and konjac gums) were mixed at a fixed binders-solid ratio of 5 % (w/w). Control (urea only) and binary mixture (urea + binder) compacts of around 300 mg were produced at different compaction pressures (50 to 200 MPa) using a punch-die set of a standard machine testing system at ambient conditions. Total porosity, apparent diffusion, and dynamic contact angle profiles of the compacts were measured. The dissolution behavior relied on the Weibull model to compare the shape, b, and scale, a, parameters. The diffusion coefficients of compacts using HPMC or gums were significantly higher than the other systems at 100 and 150 MPa. For the gum formulations, this is because of the primary particle shape distributions of the powder gums used as binders, which led to distinct compaction behavior and higher void space in the internal structure of the composites. Additionally, the higher hydrophilicity of formulations with gums also explained the rapid liquid transport through the porous compacts, with contact angle hysteresis of 14.91 ± 3.36° against 26.69 ± 3.36° for the control systems at 200 MPa. The superior hydrophilicity along with the significant hydration of the compacts using gums as binders correlated to faster swelling kinetics and superior gel strength; factors necessary to ensure a better-controlled release pattern of nutrients to the solvent phase. These results will aid in the design of controlled-release fertilizer for large scale crop operations.

627918 A Moving Porous Media Model for Simulations of Continuous Spatial Particle ALD

Abstract:

Continuous spatial particle atomic-layer deposition (ALD) reactors contain billions of fine and ultrafine particles under typical operating conditions. Simulating micron and submicron powders in a reactor with gas dosing zones on the order of centimeters presents a significant modeling challenge. However, experiments and CFD-DEM simulations in MFIX have revealed that the powder bed moves as a solid plug under typical vibratory convection conditions, so we decided to take advantage of the low relative motion between particles by approximating the powder bed as a porous media. A moving porous media model, which utilizes the sliding and layering dynamic mesh capabilities in ANSYS Fluent, was developed to enable efficient simulations of reactor-scale multiphase flow in the continuous vibrating beds used for particle ALD. Treating the packed powder bed as a solid skeleton with an infiltrating gas phase circumvents the computational expense associated with resolving particle-particle collisions. Simulations revealed the effects of packed bed and porous baseplate permeability, vibration intensity, and precursor/purge gas velocity ratios on the flow behavior inside continuous spatial particle ALD reactors. Understanding and predicting the reactor-scale behavior enables process optimization under different reactor conditions.

628439 *What Is More Important for Improved Drug Dissolution: Agglomerate Size Reduction or Enhancing Particle Surface Wettability?*

Abstract:

A prevalent challenge with fine powders, especially those below 30μm, is high cohesion that leads to flow and handling problems and more importantly, causes significant particle agglomeration that delays their dissolution rate[1,2]. Dry coating of fine powders with silica nanoparticles has been shown to be very effective in reducing cohesion and consequently may reduce the extent of agglomeration[3,4]. For poorly water-soluble particles, the reduction in agglomeration greatly enhances the available surface area, which could have a positive impact on their dissolution. However, a major concern with the dry coating is that usually hydrophobic silica particles have been found to be more effective in reducing cohesion and particle agglomeration[3,5,6,7]. An interesting question arises as to how effective the hydrophilic silica coating for agglomerate size reduction is. Even more interesting question is if the use of hydrophobic silica may lead to a greater extent of cohesion reduction, what is the adverse impact on dissolution is of the extent of increased surface hydrophobicity of drug particles that may be hydrophobic, to begin with. These two questions have largely been ignored in the bulk of the literature on dry coating. Fortunately, in a recent work, Kim et al. 2021 explored the competing relationship between those two impacts on the dissolution rate of dry coated poorly water-soluble and cohesive Active Pharmaceutical ingredients (API)[8]. Quantification of the agglomerate size and surface hydrophobicity of cohesive and micronized poorly water-soluble APIs before and after dry coating was examined in detail. As a positive surprise, the outcomes demonstrated that when the reduction in agglomerate size is effective, the dissolution is not solely dominated by the surface hydrophobicity, but it was the result of the competition between surface hydrophobicity and reduction in agglomerate size. As a corollary, when the reduction in agglomeration or the interparticle cohesion force is not sufficient, surface hydrophobicity was the main driver in the dissolution process[8]. Thus, important insights in understanding a complex phenomenon were gained, opening additional questions such as the reported observation holds for drug materials that have even higher hydrophobicity and even lower solubility in water. Higher hydrophobicity could have confounding effects in various aspects of the dissolution of agglomerated particles. For example, the study of instantization of granulated milk powders in water has pointed out multiple interacting phenomena, such as the surface wetting, sinking, penetration in the pores, agglomeration dispersion, and particle dissolution all affected by the particle's inherent solubility, its surface properties, and its size[9].

In the extended work that is the subject of the poster presentation, three different drug materials were considered while keeping their sizes comparable. To better analyze the effect of disparate water solubility, a small amount of wetting agent was added to the dissolution buffer as is the common practice for *in vitro* dissolution testing. Specifically, 0.012M of sodium dodecyl sulfate (SDS) solution[10] was added and the extent of agglomerate sizes, as well as surface wettability and dissolution rate of uncoated and dry coated model APIs, were evaluated[11,12]. The model APIs are all BCS II drugs, which were ibuprofen, griseofulvin, and fenofibrate, all having 10 microns in d_{50} were considered. In addition, for the sake of comparison with previous work[3,8], for ibuprofen, another size of 20 microns d_{50}, was also included in this study. These model APIs were dry coated with either hydrophilic (A200), or hydrophobic (R972P) fumed silica via a high energy mixer, LabRAM, at varying ranges of theoretical surface area coverage (SAC) by silica. Agglomerated and primary particle sizes were evaluated by two different particle sizers, a gravity-driven dispersion method, and a compressed air dispersion method, respectively. Following USP <711> protocol, USP IV dissolution testing was conducted in 0.012M SDS solution. Surface hydrophilicity was measured via the modified Washburn method[11,12]. The results further verified the observation reported in the earlier work by demonstrating that irrespective of the chemical nature or log P value of the targeted poorly water-soluble API, it is the extent of deagglomeration that has larger influence on how the dissolution process takes its course. In fact, that also explains why dissolution improves in certain cases of fine APIs dry coated with hydrophobic silica. In summary, this innovative study helps answer the question if and when even hydrophobic silica coating can help improve the dissolution rate of poorly water-soluble drugs based on a better understanding of the relative impacts of the surface hydrophobicity and the agglomerate size.

References

1. Yang, J.; Sliva, A.; Banerjee, A.; Davé, R. N.; Pfeffer, R., Dry particle coating for improving the flowability of cohesive powders. *Powder Technology* **2005,** *158* (1-3), 21-33.

2. de Villiers, M. M., Influence of agglomeration of cohesive particles on the dissolution behaviour of furosemide powder. *International Journal of Pharmaceutics* **1996,** *136* (1-2), 175-179.

3. Han, X.; Ghoroi, C.; To, D.; Chen, Y.; Davé, R., Simultaneous micronization and surface modification for improvement of flow and dissolution of drug particles. *International Journal of Pharmaceutics* **2011,** *415* (1-2), 185-195.

4. Han, X.; Ghoroi, C.; Davé, R., Dry coating of micronized API powders for improved dissolution of directly compacted tablets with high drug loading. *International Journal of Pharmaceutics* **2013,** *442* (1-2), 74-85.

5. Chen, L.; Ding, X.; He, Z.; Huang, Z.; Kunnath, K. T.; Zheng, K.; Davé, R., Surface engineered excipients: I. improved functional properties of fine grade microcrystalline cellulose. *International Journal of Pharmaceutics* **2018,** *536* (1), 127-137.

6. Jallo, L. J.; Ghoroi, C.; Gurumurthy, L.; Patel, U.; Davé, R., Improvement of flow and bulk density of pharmaceutical powders using surface modification. *International Journal of Pharmaceutics* **2012,** *423* (2), 213-225.

7. Huang, Z.; Xiong, W.; Kunnath, K.; Bhaumik, S.; Davé, R., Improved blend content uniformity via dry particle coating of micronized drug powders. *European Journal of Pharmaceutical Sciences* **2017,** *104*, 344-355.

8. Kim, S.; Bilgilig, E.; Davé, R.N., Impact of hydrophobicity of dry coating material on the dissolution of micronized BCS class II drug powders. *Submitted to Interntional Journal of Pharmaceutics* **2021**.

9. Schubert, H. 'Instantization of powdered food products', *International chemical engineering* **1993,** 33(1): 28-45

10. Lu, Y.; Tang, N.; Lian, R.; Qi, J.; Wu, W., Understanding the relationship between wettability and dissolution of solid dispersion. *International Journal of Pharmaceutics* **2014**, 465 (1-2), 25-31.

11. Washburn, E. W., The dynamics of capillary flow. *Physical Review* **1921,** *17* (3), 273-283.

12. Thakker, M.; Karde, V.; Shah, D. O.; Shukla, P.; Ghoroi, C., Wettability measurement apparatus for porous material using the modified Washburn method. *Measurement Science and Technology* **2013,** *24* (12).

630919 Radiation Heat Transfer in a Rotary Drum

Abstract:

Rotary drums deliver high heat and mass transfer rates, making them extensively useful for powder mixing and heating processes in metallurgical, cement, mining, pharmaceutical, detergent and other particulate processing applications. However, these processes are very difficult to model and operate because of the complex behavior affected by the particle parameters and process conditions such as particle size and distribution, shape, composition, fill level, rotation rate, and operating temperature. The current experimental research is focused on understand the heating mechanisms by conduction and radiation at high temperatures (>600ºC) and studying the effects of particle size distribution, fill level, and relatively low rotation rate (<10rpm). This work also serves to bridge the gap in the literature for experimental data and validating simulation-based research.

Experiments are performed using 2 mm diameter silica beads to investigate the granular flow and heat transfer by conduction and radiation inside a 6-inch diameter and 3-inch-long stainless-steel rotary drum. The drum is attached to two 10-inch titanium wheels, hence preventing the direct contact of the drum wall with the rollers. One end of the drum is closed using a transparent quartz window, capable of handling high temperatures. Another end of the drum is closed using a sapphire window, compatible with an IR camera, specifically chosen to obtain a high transmittance to the infrared light. The drum walls are maintained at a desired high temperature using NI-Cr resistive wires and a PID controller. The particle bed temperature evolution is recorded on the IR camera, which is used to determine the average particle bed temperature over time and calculate the wall-to-particle heat transfer coefficient.

631574 Effect of Dopant Modification in Iron Sulfide-Based Sulfur Carrier for Hydrogen Production from Hydrogen Sulfide in a One-Reactor Cyclic Sulfur Looping Scheme

Abstract:

Hydrogen sulfide (H_2S) is a hazardous pollutant, responsible for undesired material corrosion and catalyst poisoning. Emissions from industrial activities, especially in the oil and gas processing industries primarily contribute towards H_2S generation. Claus process is a widely adopted technology that decomposes H_2S into sulfur and steam. However, it is constrained by various drawbacks such as reduced sulfur recovery due to unfavorable thermodynamics, huge capital cost, and low energy efficiency. More importantly, the process consumes H content of H_2S into low-value steam due to its oxidative chemistry.

In this work, a one-reactor cyclic sulfur looping scheme is developed for decomposition of H_2S into valuable hydrogen (H_2) and elemental sulfur (S) using iron sulfide (FeS) as a sulfur carrier. It consists of two sub-steps—sulfidation and regeneration wherein FeS simultaneously decomposes H_2S into H_2, producing sulfur rich phase–FeS_x (where, x>1) and the captured sulfur is then removed during regeneration step by decomposing FeS_x into FeS in inert atmosphere. The scheme significantly reduces the requirement of processing units and overall energy requirement as compared to the Claus process. To improve upon the low reactivity of the sulfur carrier, FeS is modified by incorporating a low percentage (2%) of molybdenum (Mo) dopant which induces electronic structural changes while preserving phase integrity of FeS. The reactivity and recyclability performance of the sulfur carriers is tested using thermogravimetric analyzer and fixed bed setup and solid characterization techniques such as X-ray diffraction (XRD) and Scanning electron microscope (SEM) are used to understand the changes in the solid phases. With dopant modification, a dramatic increase of ~24% in sulfur uptake is obtained as compared to undoped sulfur carriers. Density functional theory (DFT) calculations are performed to investigate the reaction pathways, energetics, and electronic structures. Results from DFT pointed out that surface hydrogen diffusion is the rate-determining step for sulfidation of FeS, which can be significantly promoted through Mo dopant modification. Our findings demonstrate a novel approach for H_2S utilization and guide the future sulfur carrier design.

631799 Mastering the Processing Methods of Engineered Particles

Abstract:

Particle Engineering is critical in various industries and end uses such as 3D Printing, Pharmaceuticals, Agrochemicals, Functional Foods, Advanced Ceramics, Battery Technology and Cosmetics. The specific design and structure of particles give rise to their eventual properties and suitability for end use. This poster will highlight several processes and approaches we have developed at AVEKA in designing and engineering particles. Processes such as spray drying, bead formation, particle coating, and microencapsulation are all used to create designer engineered particles. Methods for controlling particle size and shape, techniques to engineer a microencapsulated particle, and methods to surface modify particles to enhance performance will be described. The poster will describe a unique underwater prilling process to form precise large round wax beads. Also discussed will be methods to attach small particles robustly onto larger particles for enhanced flow. Advanced ideas mimicking the complexities found in nature to design microencapsulated particles carrying oxygen sensitive omega three oils and other actives have also been developed.

632792 Microstructure Design and Release Kinetics of Layer-Wise Agglomerated Granules

Abstract:

Internal density plays a critical role in differentiating the capillary and disintegration kinetics in granulated systems. Pore size distributions, pore channel tortuosity, and mechanical properties dictate the final granule microstructure. Moreover, formulation functionality may also influence the structural evolution underwater. Consequently, process and formulation design significantly affect liquid penetration and dissolution onset. Studies have shown marked changes in the release profile of active ingredients attributable to optimizing composite density arrays. One way to promote material consolidation and tailor product quality attributes is through layer-wise agglomeration techniques that produce core-shell structures with a core denser than the layer. This work investigates the influence of formulation and process design on the internal structure and nutrient release of layer-wise agglomerated urea for the controlled release of fertilizers. Controlled release technology, which consists of granulating urea with polymers, is an alternative to mitigate the environmental problems derived from urea byproduct leaching in soil. Yet, there are still challenges in controlling nutrient release rates when incorporating biopolymers in the fertilizer granules. In this study, urea and a binder, xanthan/konjac gum, in powder form were mixed at a fixed binder-solid ratio to produce highly compacted composites. The composites were milled to adjusted particle sizes and subjected to a layer-agglomeration process to produce core-shell granules. The final product was dried at 80 ºC for 3 hours. The layer consisted of a mixture of urea and binder that promoted further densification of the granule structure. The core and layered granules were analyzed for their total and internal pore structure, dissolution, and microscopic behavior underwater and compared to control groups. Preliminary results for the water dissolution studies indicate that core-shell granules reduce nutrient release up to 77 % compared to core granules for the first 60 minutes of analysis. The core and layered granules containing the binder can extend the release profile in water dissolution studies for a time factor of 6 compared to the control group (with only urea). The synergism between granule density and binder functionality maintains the composite integrity while enabling binder migration and gelation at the solid-liquid interface. The gelation reduces the composite's solvation and dissolution and delays mass transport of the nutrient to the solvent phase. This study intends to unveil the role of granule internal structure in the liquid penetration, disintegration, and dissolution kinetics for a better design of controlled-release fertilizers.

636204 Analysis of Structure Formation and Layer Build-up from Dried Deposited Nanoparticle Suspension Droplets

Abstract:

The production of coating layers through deposition and convective drying of colloidal droplets on a solid surface is a procedure that is utilized in a great number of technical applications. Hereby, the solids first form individual structures, driven by the internal flows of each droplet [1]. Then, in the course of the application of more droplets, these structures subsequently merge into a connected coating layer. This incremental layer formation process, that can be seen for example during spray fluidized bed coating, showed a strong dependence of the resulting layers on the drying conditions in previous experiments [2].

In the course of this project, the incremental layer build-up from silicon dioxide nanoparticles was mimicked within a laboratory scale drying chamber. This was done by applying droplets of various sizes from the micro- to nanoliter range on a flat glass surface. After application the sessile droplets were dried under varying conditions up to technical and the resulting layer structures were analyzed in a droplet-by-droplet manner. The porosity as well as layer and droplet residue structure were measured incrementally during layer build-up by means of white-light interferometry and X-ray micro-computed tomography, in search for specific correlations between the various drying parameters and layer porosity or morphology.

Through the conducted experiments, the single deposit coffee-ring structures described by Deegan et al. [3] were identified to also be a recurring and determining phenomenon in the layer build-up process of nanoparticles. The deposits emerging at the start are made up of several ring-like formations stacked over one another or positioned next to another. In the further course of the incremental process, these structures merge via formation of solid bridges in between. This eventually results in a grown coating layer with a uniform morphology. Regarding layer porosity, it could be observed that a faster drying process, i. e. higher drying airflow velocity or temperature, results in higher porosity values. Additionally, an increase of porosity alongside an increase of deposited droplets was detected. This was determined to be due to the frequent inclusion of cavities within the resulting layer.

[1] R. Bhardwaj, X. Fang, D. Attinger (2009): Pattern formation during the evaporation of a colloidal nanoliter drop: A numerical and experimental study. New Journal of Physics 11, 075020.

[2] C. Rieck, T. Hoffmann, A. Bück, M. Peglow, E. Tsotsas (2015): Influence of drying conditions on layer porosity in fluidized bed spray granulation. Powder Technology 272, 120–131.

[3] R. D. Deegan, O. Bakajin, T. F. Dupont, G. Huber, S. R. Nagel, T. A. Witten (2000): Contact line deposits in an evaporating drop. Physical Review E: Statistical Physics, Plasmas, Fluids, and Related Interdisciplinary Topics 62, 756–765.

631710 A Deep Learning Approach for Chord Length Distribution Modeling of Two-Dimensional Crystals

Abstract:

Despite significant advances in process analytical technology, online monitoring of particle size distributions (PSDs) in industrial crystallizations remains a challenge. Laser backscattering using focused beam reflectance measurement (*FBRM*) has been widely shown to be a promising technique for online, but indirect measurement of PSDs [1]. In fact, FBRM measures chord length distribution (CLD), which does not provide a physical meaning of crystal size (e.g., [2,3]).

Towards converting CLDs to PSDs to enhance process monitoring capabilities, in this talk, we will present a deep learning approach for automatically converting simulated PSDs to CLDs which resemble the ones measured by FBRM. Deep learning and deep learning inspired methods have been recently applied to map the relationship between CLD and PSD [2,3,4], though without fully exploring their potential and limitations. In particular, we use convolutional neural networks (CNNs) [5], which are frequently used for image classification as they excel at identifying spatial patterns and relationships within data. By discretizing the PSD into finite bins, the multi-variate PSD can be effectively treated as an "image". As a result, each PSD is converted into an image and serves as the input to the CNN, while conveying the information about the frequency, position, and spacing of each bin within the discretized distribution. We use Bayesian optimization (BO) [6] to automatically tune hyperparameters of the CNN model. Specifically, BO is used to determine the optimal number of filters to use in each of the convolutional layers, as well as the optimal CNN training parameters.

To demonstrate the proposed deep learning approach, we first use simulation data of CLDs and PSDs for needle-like crystals to train and validate a CNN model. We show that the CNN model performs with an industrially pragmatic accuracy. We then investigate the generalizability of the CNN using transfer learning techniques [7], with the goal of obtaining better predictions for "unseen" types of crystals not included in the training data, or in cases where there is a limited amount of data. In particular, we examine the transfer learning by using a CNN that is trained for needle-like crystals, the dense layer of which is then retrained using a limited dataset of platelet-like crystals, while the convolutional layers part that performs feature extraction remains "frozen". We demonstrate the effectiveness of the CNN model augmented with transfer learning using experimental data. The proposed deep learning approach for converting PSDs to CLDs, which relies only minimally on system specific measurements, often time-consuming and expensive, and can be particularly useful for enhancing process monitoring capabilities in industrial crystallization processes, as well as laboratory studies for elucidating crystallization kinetics.

[1] Simon, L. L., Pataki, H., Marosi, G., Meemken, F., Hungerbühler, K., Baiker, A. & Chiu, M. S. (2015). Assessment of recent process analytical technology (PAT) trends: a multiauthor review. *Organic Process Research & Development*, *19*(1), 3-62.

[2] Szilagyi, B., & Nagy, Z. K. (2018). Aspect ratio distribution and chord length distribution driven modeling of crystallization of two-dimensional crystals for real-time model-based applications. *Crystal Growth & Design*, 18(9), 5311-5321.

[3] Crestani, C. E., Bernardo, A., Costa, C. B. B., & Giulietti, M. (2021). An artificial neural network model applied to convert sucrose chord length distributions into particle size distributions. *Powder*

Technology, *384*, 186-194.

[4] Irizarry R., Chen A., Crawford R., Codan L., & Jöchen S. (2017). Data-driven model and model paradigm to predict 1D and 2D particle size distribution from measured chord-length distribution. *Chemical Engineering Science*, 164, 202-218.

[5] Bengio, Y., Goodfellow, I., & Courville, A. (2017). *Deep learning* (Vol. 1), Massachusetts, USA: MIT press.

[6] Snoek, J., Larochelle, H., & Adams, R. P. (2012). Practical Bayesian optimization of machine learning algorithms, *arXiv preprint arXiv:1206.2944*

[7] Tan, C., Sun, F., Kong, T., Zhang, W., Yang, C., & Liu, C. (2018). A survey on deep transfer learning, "*International conference on artificial neural networks*" (pp. 270-279). Springer, Cham.

622939 Handling and Solid-State Method Development for Improved Isolation of Molecules with Complex API Structures

Abstract:

The pharmaceutical industry in general will continually encounter larger and more complex molecules due to the need for more novel disease targets. With these highly flexible and larger molecular structures come a higher propensity for solid forms of the API with more atypical behavior, e.g. variable hydrates/solvates and pseudo-stable amorphous forms that display a continuum of energetic states, rather than discrete solid form transitions. New approaches and methodologies must be explored and used to better characterize and understand these molecules to ensure robust and consistent isolation of the desired form, and to inform regulatory authorities upon submission. Pharmaceutical case studies are presented describing how adapted methodologies are used to provide better understanding of complex solid forms with regard to solid-state properties and the isolation process. The utility of different advanced solid-state characterization tools such as VH/T-XRPD, TGA-MS, perfusion calorimetry, and Raman spectrometry are reviewed and discussed. In the first case study, the aforementioned characterization tools were used to guide development of a crystallization and isolation process for a variable hydrate/solvate that exhibited small subtle changes in crystallinity at different environments, which impacted the crystallization design and performance. The second case study describes an API that undergoes form change and partial amorphization during drying. The developed solid-state methods were able to quantify the extent of amorphous content and how it is directly related to desolvation behavior. The knowledge generated by these case studies not only supported advancement through development, but will also contribute to an effective presentation of the properties of these more complex forms to regulatory authorities.

AIChE 2021 abstract for verbal presentation

Title: **Hollow crystal formation through a novel ripening mechanism in crystallization**

Authors: Fredrik L. Nordstrom*, Yongjian Wang, Huayu Li, Michelle Raikes, Brian Linehan, John Robson

Affiliation: Material and Analytical Sciences, Boehringer-Ingelheim, Ridgefield, CT

*Email: fredrik.nordstrom@boehringer-ingelheim.com

Abstract:

A novel ripening mechanism in crystallization is presented that allows for crystals to undergo morphological changes over time at seemingly constant conditions. This ripening phenomenon is different from Ostwald ripening in terms of its thermodynamic driving force and applicability across different particle sizes. Salicylic acid was crystallized in the presence of simulated structurally similar impurities, which resulted in formation of solid solutions. The impurity entrapment was elevated initially as a result of secondary nucleation and decreased during the course of the crystallization. The final crystals were thus comprised of impurity gradients having dirty cores and cleaner outer edges. Confocal Raman imaging was used to confirm the variable impurity entrapment in individual crystals showing enrichment in the center and interior of the crystals. As higher impurity levels corresponded to elevated solubility, dissolution took place from the impure center of the crystals when the solution concentration approached the solubility of pure salicylic acid. This in turn led to formation of macrotubular crystals with clearly distinguishable holes and cavities. The ripening mechanism is demonstrated experimentally at different stages in a crystallization relevant to a typical pharmaceutical isolation process. The results show not only that hollow crystals can be designed and challenging impurities can be better rejected, but more importantly it demonstrates the fundamental and dynamic interplay between impurity entrapment, solubility and desaturation kinetics that takes place throughout a real crystallization process.

Fig 1. Hollow crystal of salicylic acid after ripening

623775 Crystallization Engineering for the Robust Manufacturing of Islatravir Using SINTAX (shear-induced nucleation and thermal annealing crystallization)

Abstract:

Islatravir (MK-8591) is a potent nucleoside reverse transcriptase translocation inhibitor in development for HIV treatment and prevention. We have developed a robust crystallization method that ensures the active pharmaceutical ingredient (API) meets the required purity and physical property attributes to enable sub-milligram doses and multiple formulation types. Our crystallization takes a crude API, containing high levels of biologic and chemical impurities, and produces an API with high purity for drug substance quality and consistent particle size for content uniformity and improved manufacturing.

Herein, we highlight the development and scale-up of a shear-induced nucleation and thermal annealing crystallization (SINTAX) that delivers islatravir with a controlled particle size distribution and impurity profile[1]. We emphasize lessons learned by applying SINTAX at the lab, pilot, and commercial scales. Across manufacturing scales, we demonstrate the robustness of SINTAX and the importance of solubility models and process analytical technologies (PAT). Altogether, we show the importance of crystallization engineering to control impurity rejection, particle characteristics, and processability for the development of a life-changing drug.

References:

1. Sirota, E., Kwok, T., et al. "Crystallization Process Development for the Final Step of the Biocatalytic Synthesis of Islatravir: Comprehensive Crystal Engineering for a Low-Dose Drug." *Organic Process Research & Development* 25.2 (2021): 308-317.

629233 Evaluation of Liquid-Liquid Phase Separation and Spinodal Boundaries Using a Continuous-Flow Microfluidic Mixer

Abstract:

Liquid-liquid phase separation (LLPS), also known as oiling out, refers to the process where the liquid droplets with high concentration of solute phase separate in a supersaturated solution. Oiling-out is a common phenomenon among hydrophobic pharmaceutical compounds. The spontaneous nature of the transformation of oil droplets into solids makes LLPS very hard to measure. As a result, it is considered an undesired phenomenon and is typically avoided during the crystallization. Therefore, the most critical step in process development is to obtain the phase diagram to ensure the crystallization does not occur within the LLPS zone. Here, we have used the previously designed continuous-flow microfluidic mixer to evaluate the LLPS phenomena of Beta-Alanine. In this study, we present the isothermal ternary phase diagram of beta-alanine in water and isopropyl alcohol (IPA) or ethanol.

All the following experiments are performed using an optical microscope to track and record the appearance and disappearance of oil droplets. The microfluidic mixer consists of four tangential inlets. Two of them are streams of a concentrated beta-alanine solution in water, one is pure solvent (water), and the last one is pure anti-solvent (IPA or Ethanol). The procedure will start by fixing the ratio of the anti-solvent to solvent stream. Then we will start with the highest flowrate for the pure solvent stream at a fixed ratio of anti-solvent. After that, the flow rate of beta-alanine streams is gradually increased while that for the pure solvent is gradually decreased. The composition at which the droplets appear in the mixture is considered the spinodal point. Once the oil droplets appear, the flowrate of solvent stream is gradually increased until the droplets disappear. This composition indicates the binode. Finally, these spinodal and binodal points are used to plot the ternary phase diagram to highlight the LLPS zone.

630606 Molecular Simulations Reveal Underlying Mechanism of Cooling and Antisolvent Crystallization to Predict the Polymorphism and Growth of Organic Crystals

Abstract:

Understanding the self-assembly of molecules driven by increased chemical potential is a fundamental aspect of materials science. Crystallization of organic molecules is one such process where the increase in supersaturation is used to obtain the crystal of the desired molecule. However, the highly stochastic nature of molecular motion and the presence of multiple energy minima in the crystal energy landscape makes it difficult to predict the outcome of crystallization. This problem is more significant for the pharmaceutical industries, where multiple energy minima in the crystal energy landscape lead to polymorphism. To understand the molecular events which lead to polymorphism, we have employed molecular simulations of cooling and antisolvent crystallization processes. The simulations reveal that the rapid exchange of solvent molecules in the solvation shell and the configuration of the solvent molecules around different functional groups of the organic molecule can be used to predict the polymorphic form obtained at the end of the crystallization. In cooling crystallization simulations of glutamic acid molecules, the density of the solvent molecule in the solvation shell influences the polymorph formation. However, in the case of antisolvent crystallization of Histidine in the mixture of Water and Ethanol molecules, the rupture of the solvation shell due to the interaction with the Ethanol molecule influences the polymorphism. Furthermore, knowledge of per-molecule energy contribution in the solvation shell allows calculating the growth rates of the crystals, which can be used to control the morphology of the crystals. Overall, the knowledge of solvation shell dynamics explains and unifies the traditional empirical observations and opens up possibilities for better control strategies for crystallization.

623812 Cephalexin and Amoxicillin Crystal Shape Modification By Manipulating the Supersaturation and Wet Milling

Abstract:

Cephalexin and amoxicillin belong to the family of beta-lactam antibiotics and are among the most widely used small-molecule drugs. Both compounds are sparingly soluble in water, the solvent in their enzymatic synthesis process, and form hydrated crystals under typical synthesis conditions [1]. Unfortunately, both cephalexin and amoxicillin crystallization processes are reported to produce needle-like crystals with high aspect ratios, which can lead to significant agglomeration, long filtration times, and poor downstream processability [2]. There have been several efforts to address the above issue and to modify the crystal habit and shape in these systems. For example, it has been suggested that seeding and controlling the supersaturation generation through controlling either the cooling rate [3], synthesis reaction rate [4], or feeding rate [5] can be used to improve the shape of cephalexin monohydrate crystals.

In the present work, we examine the impact of wet milling on the crystal shape, aspect ratio, and size distribution in cephalexin and amoxicillin crystallization processes. Wet milling is commonly used to reduce the crystal size in different processes and to tighten the size distribution. Our results suggest that wet milling not only reduces the mean size of cephalexin monohydrate crystals, but it also leads to further improvements in crystal aspect ratio during the enzymatic synthesis and crystallization process. Based on a systematic study of different milling parameters, we will discuss the impact of milling on crystals of both cephalexin and amoxicillin. Furthermore, we investigate potential improvements in the aspect ratio of the final crystals by combining the milling with a series of pH manipulation steps where small, needle-like crystals are dissolved by periodically increasing the system pH (leading to higher solubility), giving a way to produce larger crystals with improved aspect ratios (Figs. 1A and 1B). The impact of the improved aspect ratio on the filtration rate of both crystals will also be discussed (Fig. 1C).

Fig. 1. Example of improvement in the aspect ratio of cephalexin monohydrate crystals by applying a series of pH swings (A), or periodic milling (B). Both cases lead to the formation of more platelike crystals compared to initially needle-like crystals. Example of the impact of aspect ratio (or size) on filtration rate (C).

Cited literature

[1] McDonald, Matthew A., Grant D. Marshall, Andreas S. Bommarius, Martha A. Grover, and Ronald W. Rousseau. "Crystallization kinetics of cephalexin monohydrate in the presence of cephalexin precursors." *Crystal Growth & Design* 19, no. 9 (2019): 5065-5074.

[2] Perini, Giulio, Carlos Avendaño, William Hicks, Anna R. Parsons, and Thomas Vetter. "Predicting filtration of needle-like crystals: A Monte Carlo simulation study of polydisperse packings of spherocylinders." *Chemical Engineering Science* 230 (2021): 116151.

[3] Ostergaard, Iben, Heidi Lopez de Diego, and Haiyan Qu. "Crystallization of cephradine polymorphs and hydrates from mixed solvents of methanol and water." *Chemical Engineering & Technology* 42, no. 7 (2019): 1512-1518.

[4] McDonald, Matthew A., Grant D. Marshall, Salami, Hossein, Andreas S. Bommarius, Martha A. Grover, and Ronald W. Rousseau. "Crystallization Optimization By Enzymatic Control of a Reactive Crystallization". AIChE annual meeting, San Francisco (2020)

[5] Li, Mingchen, Zeren Shang, and Baohong Hou. "Optimizing the Aspect Ratio of Cephalexin in Reactive Crystallization by Controlling Supersaturation and Seeding Policy." *Transactions of Tianjin University* 25, no. 4 (2019): 348-356.

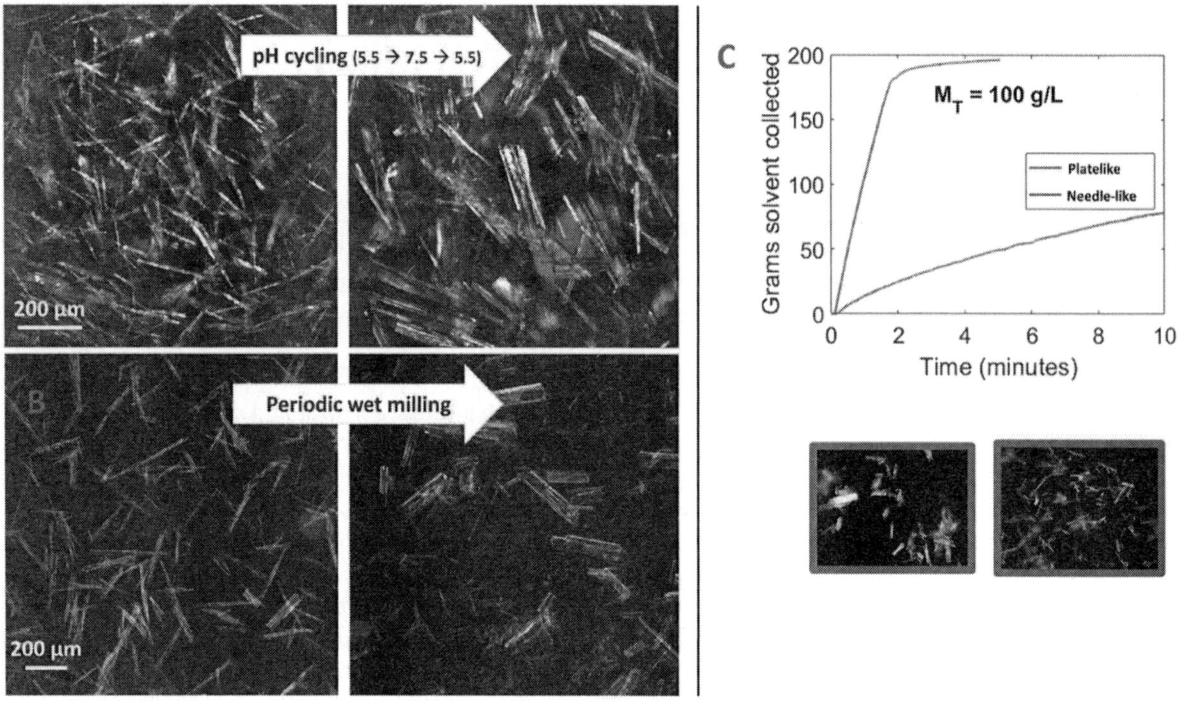
A
pH cycling (5.5 → 7.5 → 5.5)
200 µm
B
Periodic wet milling
200 µm
C
Grams solvent collected
Time (minutes)
M_T = 100 g/L
Platelike
Needle-like

625129 Accelerated Early-Stage Enabling API Crystallization Process Development and Scale-up

Abstract:

Development of crystallization, filtration/washing, and drying processes is an essential cornerstone of delivering quality Active Pharmaceutical Ingredient (API) in the pharmaceutical industry. During early-stage development, rapid timelines and the need to generate high quality API to enable GLP toxicity studies and Phase I clinical trials necessitate rapid and targeted data rich experimentation to develop robust and scalable processes. While early-stage crystallization development often falls upon the chemists, early intervention with an engineering and scalability-focused mindset can significantly reduce the development burden later in the pipeline. Such development efforts on Compound A, and the enabling data-rich workflow leveraged to yield a robust crystallization, filtration, and drying process is detailed in this work. A rapid 6-week development period, followed by a successful scale up from gram to kilo scale over a subsequent two-week period enabled a GLP toxicity study to proceed for the early-stage Compound A.

Swift crystallization solvent selection was aided by a combination of in-silico modeling using COSMO quantum chemistry software in conjunction with experimental UHPLC based solubility measurements. Selection of a suitable solvent-antisolvent system was further informed using DynoChem®, which was leveraged to generate full solubility curves over a range of temperatures and compositions with limited experimental data. Leveraging these tools is often essential in early-stage process development due to limited material availability. A favorable solubility profile for a given composition was identified, and a simple cooling crystallization was designed to afford high yield and purity. Material lean experiments using a variety of process analytical technologies (PAT) were designed to determine the metastable zone width and explore seeding conditions (seed load and supersaturation ratio) to ensure consistent crystal form control while avoiding oiling out. A complex form landscape during filtration and drying was mapped with DSC/TGA/MS and PXRD and navigated through to ensure the physical properties of the API were preserved during filtration, washing, and drying. Carefully tailored displacement washes were developed and leveraged to avoid crystal form change. A terminal wet milling and thermocycling process was also developed to control crystal surface area in conjunction with appropriate drying process understanding to achieve target specifications in the lowest cycle time. The process was successfully scaled-up to kilo scale to enable the GLP toxicity study.

Disclosures:

All authors are employees of AbbVie and may own AbbVie stock. AbbVie sponsored and funded the study; contributed to the design; participated in the collection, analysis, and interpretation of data, and in writing, reviewing, and approval of the final publication.

630526 Development and Scale-up of a Crystallization Process for a Kinetically-Unfavorable Polymorph

Abstract:

ABSTRACT: This work describes the development and scale-up of a crystallization process to isolate a thermodynamically stable but kinetically unfavorable polymorph of a small organic agrochemical prone to oiling. The process involves simultaneous addition of solution and antisolvent into a slurry of the desired polymorph, similar to semi-batch, reverse addition antisolvent crystallization.[1] For this system, the simultaneous addition process is more robust to feed variation than a traditional antisolvent process and affords suitable polymorph control, recovery, cycle time, and purity. However, the process produces a thick, but shear-thinning slurry of fine, needle-like particles that is challenging to mix well. Mixing analysis tools, including both the Fluid Mixing Processes (FMP) Excel Design Guide and computational fluid dynamics simulations were utilized to enable successful scale-up of this mixing-sensitive process from lab to pilot plant.

[1] R. Pena, C.L. Burcham, D.J. Jarmer, D. Ramkrishna, Z.K. Nagy, "Modeling and optimization of spherical agglomeration in suspension through a coupled population balance model," *Chemical Engineering Science,* 167 pp. 66-77, 2017.

632628 Acid-Catalyzed Esterification Governs the Chain Elongation and the Oriented Attachment in Cof-5 Synthesis

Abstract:

Covalent-Organic Frameworks (COFs) have found application in a wide range of fields such as catalysis, membrane separations, adsorption, and drug delivery due to their high symmetry, porosity, crystallinity, and large surface area. However, long timescales of the synthesis, difficulties in characterization, formation of amorphous intermediate, and the limited understanding of the underlying reaction network have prevented the controlled synthesis of COF crystals. Furthermore, the existing solvothermal synthesis techniques are difficult to reproduce. To overcome these limitations, improve reproducibility, and obtain quantitative rate contributions of different polymerization phases, we employ experimental studies of a wide range of operating conditions to synthesize COF-5 crystals. The yield and grain sizes of the COF crystals were characterized using reliable and reproducible ex-situ techniques such as Fourier Transformed Infrared Spectroscopy (FTIR) and powdered X-ray diffraction (P-XRD). Optimization of the temperature, reactant concentrations, type of solvents, and catalysts have provided significant mechanistic and resulted in low reaction times, high yields, and larger particle sizes. The insights gained from the comprehensive experimental studies allowed the development of a microkinetic model which simulates millions of nodes in the COF reaction network. The error-minimized parameters in the microkinetic models reproduce the experimental results of COF-5 yield and grain sizes in a predictive manner. Furthermore, the microkinetic approach also validates the proposed reaction mechanism obtained from the experimental studies. Results suggest that the acid-catalyzed esterification governs the chain elongation and the oriented attachment of COF-5 crystals.

621619 Enabling a Selective Dissolution Scheme for the Removal of Fines in Crystallization with a Hydrocyclone: Modeling and Experimental Validation

Abstract:

Effective methods to tune the particle size distribution (PSD) are of major interest in the field of crystallization, with the aim of facilitating the downstream processing, improving the mechanical properties of the powder, and, in the case of crystalline pharmaceuticals, modulating the absorption rate of the active ingredient in the human body.

Fines, i.e., crystals of a few tenths of micrometers in size, are often an undesired portion of the PSD, as they are responsible for extended filtration times. For their removal, temperature cycling is usually applied, consisting of one or more repetitions of a partial dissolution step followed by a growth step [1,2]. Temperature cycling has the advantage of being a very simple procedure; however, depending on the PSD and on the compound-specific behavior of crystallization, its efficiency may be limited, and the outcome may not be well predictable. Two factors limit the efficiency and predictability of temperature cycling. First, the partial dissolution step is non-selective, as it acts on the whole suspension and also affects larger crystals, thus decreasing the productivity; second, the evolution of the PSD strictly depends on the kinetics of dissolution and growth, which are generally unknown, and require tedious lab-work to be evaluated.

To address both shortcomings of temperature cycling, we have proposed and analyzed in detail through modeling an alternate process featuring a hydrocyclone [3], referred to as a selective dissolution process and outlined in Figure 1a. Hydrocyclones are common and inexpensive devices employed in several fields of chemical engineering, where they can act as classifiers to separate particles based on their size [4]. In the selective dissolution process, first the hydrocyclone separates finer from coarser crystals, then the suspension containing the fines is heated, so that only the particles in the targeted size range are dissolved. Further, the separation in the hydrocyclone depends on its design and operating conditions, and on the physical properties of the suspension, thus making the process less reliant on the kinetics of crystallization.

In this contribution we present both the main findings of the above-cited modeling study and the results of novel validation experiments, highlighting the trends and trade-offs that characterize the operation of a hydrocyclone and their effect on the overall performance of the crystallization process. Particular attention is devoted to process design, and a pathway for the selection of a suitable hydrocyclone design for a given target of fines removal is outlined.

The experimental campaign has been performed with β L-glutamic acid in water. 3D printing has been exploited to manufacture the hydrocyclones in-house (shown in Figure 1b), so as to enable superior flexibility in the design of units with different geometries. The experimental results demonstrate the advantage of the selective dissolution process over standard temperature cycling, and validate the modeled trends. Finally, the experimental data has been used to fine-tune the model predicting the separation performance of the hydrocyclone, which can be therefore used reliably for process design.

[1] Nagy, Z. K.; Aamir, E.; Rielly, C. D. Internal Fines Removal Using Population Balance Model Based Control of Crystal Size Distribution under Dissolution, Growth and Nucleation Mechanisms. *Growth Des.* **2011**, *11*, 2205–2219.

[2] Salvatori, F.; Mazzotti, M. Manipulation of Particle Morphology by Crystallization, Milling, and Heating Cycles: Experimental Characterization. *Eng. Chem. Res.* **2018**, *57* (45), 15522–15533.

[3] Binel, P.; Mazzotti, M. A Selective Dissolution Process Featuring a Classification Device for the Removal of Fines in Crystallization. *Eng. Chem. Res.* **2021**, *60* (1), 614–628.

[4] Svarovsky, L. Hydrocyclones. In *Solid-Liquid Separation*; Svarovsky, L., Ed.; Butterworth-Heinemann: Oxford, U.K., 2001; pp 191–245.

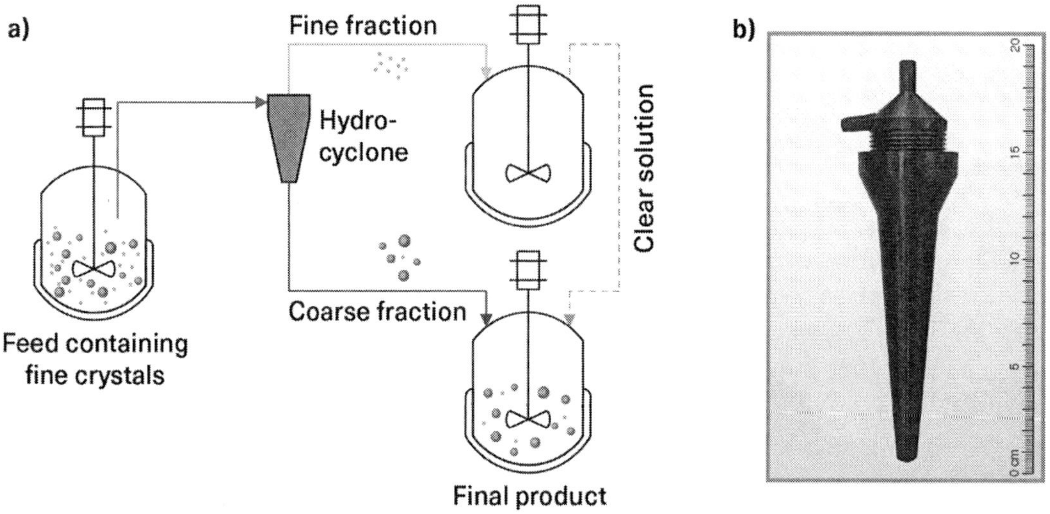

Figure 1. (a) Scheme of the selective dissolution process. The feed suspension is pumped through the hydrocyclone, which splits it into two streams; one stream contains most fine crystals while the other contains most coarse crystals. The stream containing the fines is collected in a vessel where complete dissolution occurs. The clear solution is then combined with the suspension containing the coarse crystals, and the supersaturation is depleted until equilibrium, i.e., the end of the process. (b) Photograph of a 3D printed hydrocyclone. For improved printing performance and flexibility, it is composed of a top cylindrical part and a bottom conical part. The top part is screwed into the bottom part, ensuring a tight fit and a smooth transition between the two.

627313 Crystallization Modeling of a Pharmaceutical Compound for Digital Twin Based in-Silico Optimization with Experimental Validation

Abstract:

The design of crystallization processes to achieve robust critical quality attributes (QCAs) is a practical challenge in the pharmaceutical industry. The current industrial standard still highly depends on Quality-by-Design (QbD), which requires intensive experimentation using traditional design of experiments (DoE) to find the operation conditions in the experimental space that satisfy the QCAs. With the current emerging Industry 4.0 practices the emerging paradigm has become the introduction of digital twins, which provide a predictive tool to be used in the optimization of the crystallization processes, or for in-silico DoE studies.[1]

In this work, we present the development and validation of a digital twin for the crystallization process of a pharmaceutical API (Compound A), from Takeda Pharmaceuticals International Co., which forms needle-like crystals. Several model training and validation experiments were performed with temperature cycling to promote the growth of the crystals for kinetic studies. For data acquisition and system monitoring during these experiments, process analytical technology tools (PAT) were used, including Mettler Toledo's ParticleTrack G400 for in-line chord length distribution (CLD), for detecting the crystallization events such as nucleation and dissolution, as well as for the crystal count measurement, the Zeiss MCS621 ATR-UV/Vis spectrophotometer for concentration measurement, and the Mettler Toledo's ParticleView v19 for in-situ microscope images. Additionally, Malvern's Mastersizer 3000 Hydro MV was used as an off-line characterization tool to measure crystal size distribution (CSD) and the data from the PAT tools were directly used in the parameter estimation step.

Two novel features were introduced during the parameter estimation for digital twin development. First, to make the model parameter estimation more robust, the focused beam reflectance measurement (FBRM) data was directly used in the parameter estimation without further transformations. Secondly, a novel size dependent growth (SDG) rate expression was used in the model to capture the CSD dynamics considerably better than the standard SDG rate models. The developed digital twin includes secondary nucleation, SDG, and dissolution mechanisms, which were identified by the designed training experiments and the objective that required thermocycling for the desired CQAs. The aim was to grow the crystals as large as possible without exceeding a batch time of 40 h while having product crystals with bulk density values larger than 0.25 g/mL. For this purpose and from previous experience, the training and validation experiments were designed to have temperature cycles.[2,3] The developed model was used for in silico DoE and process optimization and the simulation results were validated experimentally, demonstrating the benefits of model-based digital design for crystallization process development. In addition, the benefits of having a digital twin for in-silico optimization were demonstrated by comparing the findings from an experimental DoE work on Compound A to the ones from the optimized profiles, which were obtained by using the digital twin. The results demonstrated that the developed model was accurate and predictive enough to validate the general conclusions related to the required key features of the optimal temperature profiles identified by the experimental DoE. These results indicated that in order to obtain larger crystals with narrower CSD, multiple temperature cycles should be implemented at higher temperature ranges (50-65 °C). It was also shown that having a higher range for cycling with multiple cycles has a larger effect on product crystals size and bulk density than changing the seed type and loading.

The conclusions of the in-silico DoE and digital design were validated by experiments, demonstrating that the model was able to identify the key features of the optimal operating conditions for the crystallization of Compound A, when the aim was to produce large crystals by implementing temperature cycles. The orders of magnitude difference in the time required to reach this conclusion by in-silico and experimental DoE also demonstrated the benefits of having a digital twin of the process that can enable rapid crystallization process development by using a digital design framework.

Declaration of Competing Interest

The authors declare that they have no known competing financial interests or personal relationships that could have appeared to influence the work reported.

References:

1. Su, Q. *et al.* A perspective on Quality-by-Control (QbC) in pharmaceutical continuous manufacturing. *Comput. Chem. Eng.* **125**, 216–231 (2019).
2. Jiang, M. *et al.* Modification of crystal shape through deep temperature cycling. *Ind. Eng. Chem. Res.* **53**, 5325–5336 (2014).
3. Simone, E., Klapwijk, A. R., Wilson, C. C. & Nagy, Z. K. Investigation of the Evolution of Crystal Size and Shape during Temperature Cycling and in the Presence of a Polymeric Additive Using Combined Process Analytical Technologies. *Cryst. Growth Des.* **17**, 1695–1706 (2017).
4. Mydlarz, J. & Jones, A. G. On the estimation of size-dependent crystal growth rate functions in MSMPR crystallizers. *Chem. Eng. J.* **53**, 125–135 (1993).

630582 CFD Investigation of the Nanoparticle Distribution within the Peritoneal Cavity for the Treatment of Ovarian Cancer Metastasis

Abstract:

Introduction/Purpose:

The peritoneal cavity is a common site of metastatic disease for a variety of cancers including ovarian, gastrointestinal tract, lung and breast. It is also a site for a wide spectrum of non-neoplastic conditions. Nanoparticle-based therapeutics have become increasingly popular for i.p. delivery due to their increased peritoneal residence time and the ability of the particles to freely move throughout the entire cavity for uniform distribution of various drugs. It is, however, unknown what specific nanoparticle formulation and infusion characteristics optimally distribute these particles throughout the cavity to uniformly reach the target tumors. The main objective of this work was to advance the fundamental understanding of the mechanics of complex multiphase flow patterns, distribution and deposition of nanoparticles within the peritoneal cavity using computational fluid dynamics (CFD) and discrete phase model (DPM).

Methods:

A 3D numerical model was developed representing human torso with multiple tumors. The nanoparticles were injected inside the peritoneal cavity at a specific flow rate obtained from experiment. The fluid phase was treated as a continuum and governed by the Navier-Stokes equations, while the dispersed phase was solved by tracking a large number of particles through the calculated flow field. The model was developed using DPM model in ANSYS FLUENT software.

Results:

After running various simulations, the nanoparticle formulation properties such as nanoparticle density was investigated in order to understand the nanoparticle distribution in the peritoneal cavity. These results were then correlated with real-time *in vivo*, optical imaging of mesoporous silica nanoparticle distribution within the peritoneal cavity of mice bearing peritoneal metastasis at various time points up to 24 hours post injection.

Conclusions:

The results of this study were used to gain insight in the overall design of nanoparticle formulations for the treatment of peritoneal tumors by understanding the impact nanoparticle formulation properties on nanoparticle distribution within the peritoneal cavity. This is an undergoing study which could be helpful in studying the effect of formulation and injection properties on the nanoparticle distribution.

632084 Measurements on the Acoustic Properties of Powders

Abstract:

In previous research, the feasibility of using Helmholtz Resonance to measure fill volume of the material in a hopper and the discharge rate from it, was investigated. This is done by applying an excitation sound wave to a hopper containing a particulate material and measuring the resonance frequency of the system. It was observed that the resonance quality of such a system greatly depends on the particle properties of the sample material and the resonance frequency range of the hopper. To gain a better understanding of the correlation between the physical properties of a bulk material and the acoustic behavior of a resonator containing it, a vertical impedance tube has been constructed. The impedance tube consists of a sound generator, a tube of adjustable length, and a cylinder that holds a powder sample. The apparatus is used to determine various acoustic parameters including normal incidence sound absorption coefficients. Sound absorption coefficient was measured for different types of particles with different size distributions and bulk densities, at different material depths in the sample holder. Initial experiments have largely focused on readily available foodstuff materials and early indications on sound absorption are consistent with the trends shown in past trials with a laboratory scale resonator discharging mustard seeds and agricultural lime.

622180 A Comprehensive Review: Metal Dust Cloud Combustion & Nonstandard Explosion Protection Applications

Abstract:

Combustible dust explosions continue to present a significant threat for a variety of industries relying on pneumatic conveyance and bulk powder storage. Although industry standards offer a prescriptive approach toward hazard definition and protection requirements, deviation from such guidance exposes critical gaps in understanding and obstructs the implementation of safe explosion mitigation solutions. Beginning with a discussion of inherent metal dust explosion risks, this comprehensive review provides a consolidation of the motivations and results for a range of investigations, with the objective of illustrating unique aspects of metal dust combustion phenomena and other nonstandard application scenarios. Included here within is a technical overview of the following recent published works:

- Characterization of thermal stability and heat absorption potential for various suppression agent compositions (https://doi.org/10.1021/acs.iecr.8b06143)
- Large scale (1 m^3 combustion sphere) explosibility and suppression of iron and aluminum powder deflagrations (https://doi.org/10.1021/acs.iecr.9b04021)
- Consequence prediction for dust explosions involving adjacent interconnected vessels using FLACS modeling techniques (https://doi.org/10.1016/j.jlp.2020.104149)
- Effect of particle morphology on aluminum dust explosion sensitivity and severity (https://doi.org/10.1016/j.jlp.2021.104396)

631863 A Linear Peristaltic Device for Pumping and Mixing Liquids and Slurries

Abstract:

Mimicry of the bio-mechanical action of the soft walled human digestive tract could provide a viable method for the transportation and mixing of liquids or slurries, with potential application in situations where metallic components are not desirable, such as in the production and transport of propellant substances.

The gastrointestinal tract is nature's biological slurry pump. The actions of the muscles within the tract wall produce peristaltic waves and segmented contraction activity that efficiently mix and transport digestion materials and products. A biomimetic physical model of the gastrointestinal tract has been developed that uses controlled hoop actuators (unit constrictors) to simulate the peristaltic contraction and expansion action of the muscles which mix or propel the material within a closed (stomach) or open ended (oesophagus or duodenum) gastric conduit respectively.

A prototype linear peristaltic device based, on our gastric model, performed well in an exploratory study of pumping performance for a Newtonian fluid (water), achieving flow rates up to 365 mL/min and pumping heads up to 5.3 meters. Work is in progress to investigate the effectiveness of the device for pumping and mixing viscous slurries.

623043 Preparation of Abamectin Nanoparticles By Flash Nanoprecipitation for Extended Photostability and Sustained Pesticide Release

Abstract:

Abamectin has been widely used as a biopesticide to control pests, such as pinewood and root-knot nematodes. However, the low aqueous solubility and poor photostability of abamectin have largely hindered the delivery and bioavailability of conventional formulations, causing a significant rise in costs as well as limiting the large-scale application in crop production. Here, we report the formulation of abamectin nanopesticides for extended photostability and sustained release, using a continuous and scalable technique called Flash NanoPrecipitation (FNP). Cost-effective and biocompatible stabilizers, hypromellose acetate succinate (HPMCAS) and lecithin, were used in the formulation with optimized mass ratios for nanoparticle stability. The encapsulation efficiency of abamectin in nanoparticles was higher than 90% with a mass loading capacity up to 50%. The optimized formulations not only showed improved photostability compared with free abamectin but also achieved more sustained release behaviors in a controlled manner. These results suggest that the abamectin nanoparticles formulated by FNP can improve the performance of the aqueous solubility, photostability, and controlled release of abamectin. In addition, our work highlights the potential of FNP for the mass production of low-cost abamectin nanopesticides with affordable materials, which is essential for the large-scale application of nanopesticides in food and agriculture.

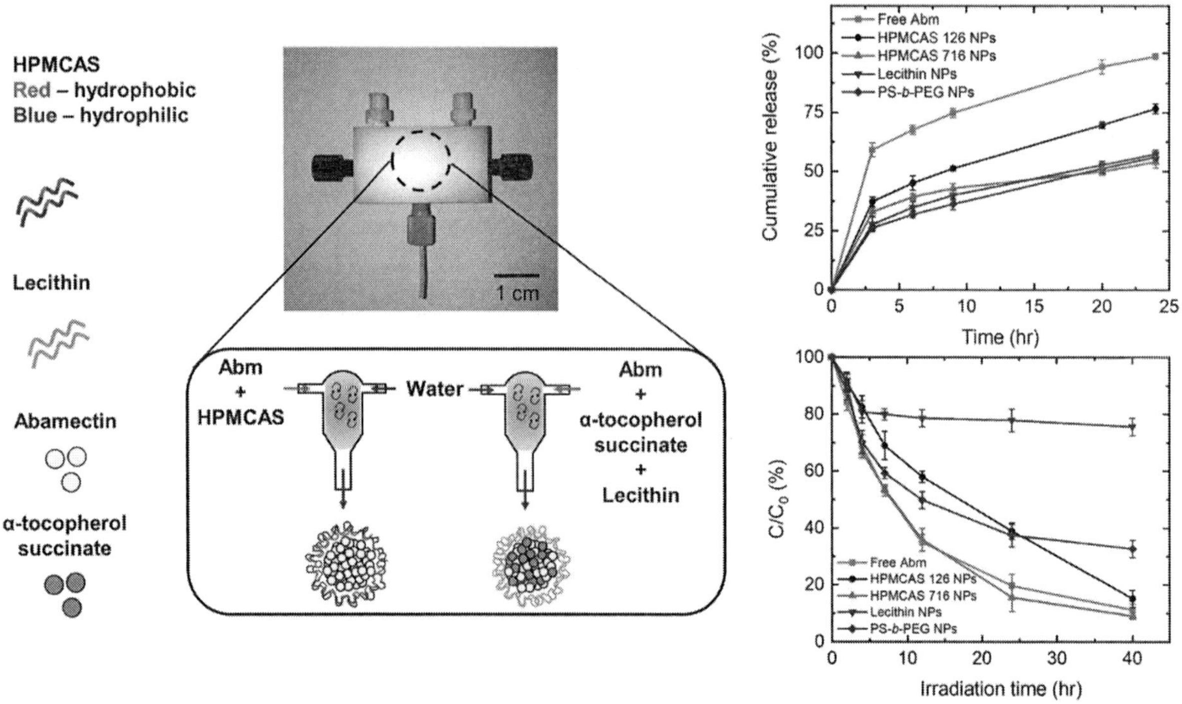

625218 Geometric Similarity on Force Evaluation for Scaled-up Particle Model in DEM

Abstract:

Discrete Element Method (DEM) are used to simulate various kinds of particulate systems. In DEM, it is straightforward to consider complex inter-particle forces such as contact, liquid bridge and cohesion forces in simulation, which may be difficult to achieve with other methods including the continuum method. One of the major challenges in DEM is the high computational cost to track the movement of all individual particles in a system. This makes it difficult or practically impossible to perform simulation of fine particles in industrial-scale systems.

The scaled-up particle model, which is sometimes referred to as the coarse grain and/or the discrete parcel models, has been increasingly popular to overcome the aforementioned problem: large particles are used in simulation to mimic the behaviour of the original small particles. The scaled-up particle model in the literature may be largely classified into the parameter scaling and the direct force scaling approaches. The former employs scaled-up physical properties and parameters (often based on some dimensionless parameters) to achieve similarity to the original particle system, whilst the latter first estimates the forces acting on the original particles using the original physical properties and variables, and then the resultant forces are directly scaled.

In the scaled-up particle model, the original particles are not explicitly simulated but represented by scaled-up particles. Therefore, particularly in the direct force scaling approach, it is important to accurately evaluate the original particle variables related to the force calculation, such as the translational velocity, angular velocity, particle overlap and inter-particle separation distance, from the scaled-up particle variables. While some correlations for translational and angular velocities are derived based on the conservation of total kinetic energy, those for overlap and inter-particle separation distance are not well-grounded in theory. However, little attention has been paid to this topic in the literature.

The present work discusses the evaluation method of original particle variables for the force calculation in the scaled-up particle model. The most notable difference from the previous work in the literature is that the overlap and inter-particle separation distance of original particle are evaluated based on geometric similarity. Simulations of uniaxial compression of a particle bed as well as adhesive/cohesive particles in a mixer are performed with the proposed approach which sees a good agreement with the original simulations in terms of the stress-strain relationship and velocity distribution.

628161 A Scale-Independent Approach to Modeling Filtered Drag in Fluidized Gas-Particle Flows By Introducing the Solid Volume Fraction at a Larger Scale

Abstract:

The filtered drag force is well accepted as the most important constitutive term in practical coarse-grid simulations of gas-solid flows. Currently, four critical issues including considering the effect of macroscale condition, reducing complexity of the model, improving the scale independence and enhancing the adaptability for different flow regimes, have been addressed in recent studies[1-4] on modeling the filtered drag correlation. To improve the performance of the filtered drag model in these aspects, a novel modeling approach has been developed. In this modeling approach, the modeling database is generated through fully resolved two-fluid model (TFM) simulations coupled with kinetic theory of granular flows in periodic sedimentation systems. To consider the effect of macroscale condition, a new marker, the filtered solid volume fraction at a larger scale whose side length is three times of the filter size, is introduced. The raw data are filtered and binned in terms of the available markers including dimensionless filter size, filtered solid volume fraction, filtered solid volume fraction at the second scale, dimensionless filtered slip velocity and dimensionless filtered gas pressure gradient. A good linear correlation between filtered gas pressure gradient and the bin-averaged filtered drag force has been obtained. And this correlation is independent of the filter size owing to introduction of the filtered solid volume fraction at large scale. Moreover, the effect of the filtered slip velocity on this correlation is only moderate at the dilute region. Based on these features, the ratio between the filtered drag force and filtered gas pressure gradient is treated as the dependent variable and the effect of the filtered slip velocity is neglected to make the modeling process concisely. With these simplifications, the models only correlated to the filtered solid volume fraction at the filter scale and that at the larger scale are developed. Subsequently, the new developed models are evaluated in terms of a priori test and a posteriori test for three typical fluidized regimes including bubbling, turbulent and fast fluidized bed. Compared with the gas pressure gradient-dependent model proposed in our previous study[5],the new developed models exhibit an obvious improvement. Additionally, the predictions of the new developed models have good agreements with experiments and the fine-grid results for different flow regimes.

Reference

[1] Y. Jiang, J. Kolehmainen, Y. Gu, Y.G. Kevrekidis, A. Ozel, S. Sundaresan, Neural-network-based filtered drag model for gas-particle flows, Powder Technol. 346 (2018) 403-413.

[2] J.H. Cloete, S. Cloete, S. Radl, S. Amini, On the choice of closure complexity in anisotropic drag closures for filtered Two Fluid Models, Chem. Eng. Sci. 207 (2019) 379-396.

[3] J. Mouallem, N. Chavez-Cussy, S.R. Niaki, C.C. Milioli, F.E. Milioli, On the effects of the flow macro-scale over meso-scale filtered parameters in gas-solid riser flows, Chem. Eng. Sci. 182(2018) 200-211.

[4] S.R. Niaki, J. Mouallem, N. Chavez-Cussy, C.C. Milioli, F.E. Milioli, Improving the accuracy of two-fluid sub-grid modeling of dense gas-solid fluidized flows, Chem. Eng. Sci. 229 (2021) 116021.

[5] M. Jiang, X. Chen, Q. Zhou, A gas pressure gradient-dependent subgrid drift velocity model for

drag prediction in fluidized gas-particle flows, AIChE J. 66 (4) (2020) e16884.

628169 Direct Numerical Simulations of Dynamic Gas-Solid Suspensions: A New Inhomogeneous Drag Model for CFD-DEM Simulations

Abstract:

The computational fluid dynamics-discrete element method (CFD-DEM) at the laboratory scale has shown to be effective in capturing the main characteristics of gas-solid flows. The accuracy of gas-solid drag models is critically important for the predictability of simulations on gas-solid fluidized beds. A number of studies have shown that the inhomogeneities have a pronounced effect on the drag in gas-solid systems [1-3]. In this work, direct numerical simulations of flows past fluidized spheres in cubic domain with full-period boundary conditions are performed to investigate the effect of inhomogeneities on the drag at low Reynolds numbers. Two powerful sub-grid quantities have been employed to quantify the extent of inhomogeneities in the literature [3]: the drift flux, measures the correlation between the solid volume fraction and the fluid velocity, and the scalar variance of solid volume fraction, determined by the particle configurations. Due to the unresolved flow structures at the sub-grid scale in CFD-DEM simulations, the drift flux cannot be obtained directly and extra closure for it is required when the drag model based on the drift flux is applied in CFD-DEM simulations. For lack of effective approach to estimate the drift flux perfectly, drag model with the modelled drift flux usually presents an unsatisfactory result in predicting the drag. In terms of the particle configuration and velocity are available in CFD-DEM simulations, a newly defined quantity named the solid drift flux, which is a measure of the correlation between the gas volume fraction and the particle phase velocity, is proposed in this work. Then a new inhomogeneous drag model is developed by taking the effect of both the scalar variance of solid volume fraction and the solid drift flux into consideration. Comparison is made between the newly proposed drag model and prior drag models [4-5]. The results show that the new drag model achieves a good performance for describing the effect of inhomogeneities on the gas-solid drag force, which can be applied in CFD-DEM simulations without any additional closures for sub-grid quantities.

Reference

1. G. Zhou, Q. Xiong, L. Wang, X. Wang, X. Ren, W. Ge. Structure-dependent drag in gas-solid flows studied with direct numerical simulation. Chemical Engineering Science. 2014;116:9-22.

2. X. Chen, N. Song, M. Jiang, T. Ma, Q. Zhou. A microscopic gas-solid drag model considering the effect of interface between dilute and dense phases. International Journal of Multiphase Flow. 2020:103266.

3. G.J. Rubinstein, A. Ozel, X. Yin, J.J. Derksen, S. Sundaresan. Lattice Boltzmann simulations of low-Reynolds-number flows past fluidized spheres: effect of inhomogeneities on the drag force. Journal of Fluid Mechanics. 2017;833:599-630.

4. D. Gidaspow. Multiphase flow and fluidization: continuum and kinetic theory descriptions: Academic press; 1994.

5. M.A. van der Hoef, R. Beetstra, J. Kuipers. Lattice-Boltzmann simulations of low-Reynolds-number flow past mono- and bidisperse arrays of spheres: results for the permeability and drag force. Journal of Fluid Mechanics. 2005;528:233-254.

630233 Analysis of Fluidization and Particle Temperature in Semicontinuous Fluid Bed Drying of Pharmaceutical Granules

Abstract:

The fluidization dynamics and spatial heterogeneity in semicontinuous drying of pharmaceutical granules are still a black box in the understanding and mechanistic modeling of this process. Therefore, this work investigates the dominating phenomena in the fluid bed dryer of a continuous pharmaceutical wet granulation line, which are acting on the fluidization regime and heat transfer between the gas and solid phase. The gained knowledge about these phenomena will, in a later phase of the research, be used to calibrate and validate a three-phase computational fluid dynamics (CFD) model. A thorough and novel experimental data collection campaign is performed to create spatial temperature profiles considering the process features. Simultaneously, the effect of the most influential process settings on the fluidization features is also evaluated. Along with that, the effect of the observed phenomena on the obtained key product quality, the residual moisture content and granule size distribution, is taken into account. In this way, additional information needed to translate pharmaceutical fluidization processes into predictive models is provided, which makes model-based process analysis for the application of continuous manufacturing in the pharmaceutical industry, more concrete.

The fluidization regime and heat transfer are investigated in a ConsiGma-1™ semi-continuous fluid bed dryer, which is fed with wet granules from a twin-screw granulator through gravimetric transport. The Perspex wall of this drying system makes it possible to accurately capture the fluidization behavior through a video recording system followed by image processing. Supplemented with several other sensors such as temperature, pressure, and relative air humidity loggers, it was possible to cover the spatial effects of the majority of the influential factors on the phenomena under investigation.

39 different drying experiments were performed with continuous filling of wet granules, which were supplemented with experiments on dry granules to improve the process understanding. 27 of the wet granule experiments were performed to characterize the fluidization regime and to link this regime with the product quality. For those experiments, the liquid to solid ratio was varied between 6.36 and 9.78 wt%, and the dryer inlet air flow rate between 50 and 84 m^3/h. The filling time of the dryer cell, governing the cell load, was varied between 60 and 180 seconds. To investigate the energy transfer between the inlet air and the drying granules, 12 experiments were performed with a formulation that was enriched with a thermo-sensitive colored component (LCR Hallcrest Ltd, Flintshire, UK). These components induce a change in color of the granules in function of their temperature, making the spatial temperature distribution in the bed observable. [1] For this set of experiments, the granulation settings were kept constant while the following drying settings were varied: the incoming air temperature between 35 and 60°C and the dryer inlet air flow rate between 50 and 66 m^3/h. The total drying time was always at least two times the filling time, ensuring that a stabilized fluid bed was achieved during the experiment.

An obvious observation can be made that, during the filling time, the fluidization regime will transition from a turbulent regime to a bubbling bed. Due to the very low cell loading and the constant inlet air flow rate, a high bed height is observed at the beginning of the loading period. This will result in efficient heat exchange between the solid and gas phase, inducing a high evaporation rate. As evaporation consumes the energy supplied by the fluidizing air, the granules' temperature does not increase significantly in the early drying stage. [2] Proportional to the filling time, the loading of the drying cell increases, which causes an increase in the total bed weight. Henceforth, the bed transitions from a turbulent to a bubbling fluidization regime. Through the reduced bed height, the contact surface between the solid and the air diminishes. Consequently, the energy exchange efficiency and average evaporation rate over the bed decrease.

However, not only does the fluidization behavior determine the evaporation rate during fluid bed drying, the newly gathered experimental data also indicates segregation effects depending on the granule size distribution. Together with the observed preferential granule flow patterns in the bubbling bed, this phenomenon results in a heterogeneous evaporation behavior of the granules. Spatially, different granule sizes and granule temperatures are observed during one experiment, more specifically the relative distance of a granule to a bubble will have a large influence on the temperature of that granule.

In addition to the new insights, this work also provides a wealth of data to calibrate and validate predictive models to, supplying an abundance of quantitative data in different conditions to build an accurate model. Specifically, the spatial nature of the data is used to provide a qualitative three-phase computational fluid dynamics (CFD) model, which will help to create a new generation of highly predictive models that capture the effect of process settings and granule and material properties on the drying dynamics.

References:

[1] Lakio, S., Heinämäki, J., & Yliruusi, J. (2010). Colorful drying. *AAPS PharmSciTech*, *11*(1), 46–53.

[2] D.M. Parikh, J.A. Bonck, M. Mogavero, Batch fluid bed granulation, in: D.M. Parikh (Ed.), *Handbook of Pharmaceutical Granulation Technology*, Marcel Dekker Inc., New York, 1997, pp. 227-302.

625809 Magnetic Drug Screening Nanoplatform Using Immobilized Ion Channels As Targets

Abstract:

Many studies have shown that voltage-gated sodium ion channel (VGSCs, Nav1.1-1.9) play an important role in cancer metastasis with certain subtypes associated with specific cancer types. For example, substantial evidences have shown Nav1.5 overexpression in metastatic breast cancer cells (*e.g.*, highly metastatic MDA-MB-231 cells). The Nav1.5 overexpression was also associated with metastasis and tumor progression *in vivo* and in clinical patient samples. Therefore, targeting ion channels is evolving as a direction for metastatic cancer therapy. Magnetic bead separation have been explored for direct compound fishing through the immobilized protein targets on the bead surfaces from a mixture. Unfortunately, magnetic bead separation suffers from significant non-specific binding of compounds to the bead surfaces, which makes them not suitable for screening complex matrices. In addition, the immobilized proteins are limited to cytosolic proteins (e.g. enzymes and antibodies). Transmembrane proteins are not suitable because they require boundary lipids for function. This is a great limitation, because over 60% of all modern pharmaceuticals target membrane proteins. We recently developed a novel drug screening platform based cell-membrane coated magnetic nanoclusters (CMMNs), where the functional transmembrane receptors on the coating membrane serve as "smart baits" for direct identification of binding compounds and the encapsulated magnetic nanoparticles enable quick isolation of identified compounds.

In this presentation, using Nav1.5 as a model system, we evaluated the potential of this drug-screening platform in identifying binders from artificial mixtures and a synthetic library containing about 100 ion channel blockers. The CMMNs were prepared using MDA-MB-231, a highly metastatic breast cancer cell line with known Nav1.5 overexpression. The formation of CMMNs was confirmed with transmission electron microscopy. The presence of Nav1.5 receptors on CMMN surfaces was studied using confocal microscopy with antibody labeling. The fishing experiments were performed by incubating CMMNs with an artificial mixture containing a known ion channel blocker and several non-binders and by incubating CMMNs with *a synthetic library* at 37 °C for 20 minutes. The identified Nav1.5-bound compounds were analyzed using a Water Xevo G2xs QTof-MS with iclass UPLC system. Our fishing experiments suggested that the immobilized Nav1.5 receptors on CMMN surfaces were able to selectively retain the known blocker and could identify compounds from the synthetic library.

638880 Thomas Baron Award in Fluid-Particle Systems (Sponsored by Shell): Design of Multiphase Reactors for the Synthesis of Renewable Chemicals and Semiconductor Nanocrystals

Abstract:

The first part of the presentation will focus on the design of fluidized bed reactors for catalytic fast pyrolysis (CFP) of lignocellulosic biomass to renewable value-added chemicals. CFP has emerged as a very promising method for single step thermal conversion of biomass to liquid fuels and chemicals. The production of aromatics from lignocellulosic biomass using ZSM-5 catalysts was studied in a laboratory-scale bubbling fluidized bed reactor using a combination of experiments and computational fluid dynamics simulations. The operating conditions of the reactor as well as the morphology and mesoporocity of the ZSM-5 catalyst were optimized to maximize the yield and selectivity in benzene, toluene, xylene (BTX) and other aromatics and to minimize catalyst deactivation due to coking. Modeling and computer simulations of the downward flow of thermally decomposing biomass particles in the vertical feed tube of the reactor enabled the development of performance diagrams that can be used to identify operating conditions leading to rapid downward suspension flow without undesirable transitions to slow moving bed flow or reversal of the gas flow in the tube. The second part of the presentation will focus on the synthesis of semiconductor nanocrystals that exhibit size-tunable fluorescence and find applications in ultra-high color definition displays, solar energy conversion, and biosensors. We have developed novel aerosol and colloidal synthesis methods for producing II-VI compound semiconductor nanocrystals that emit in the visible part of the spectrum. The most promising synthesis method exploits the dispersed nanodomains of microemulsions and liquid crystals as microscopic templates for controlling particle size, shape, and size distribution. The mechanism of particle nucleation and growth in spherical nanodomains was studied using Kinetic Monte Carlo simulations. Generalized scaling laws were obtained that describe the growth of nanoparticles by coalescence in confined domains. The doping mechanisms and thermodynamic stability of core-shell nanocrystals were studied using Density Functional Theory calculations. Reactor configurations are being developed that are suitable for large-scale modular manufacturing of nanocrystals with narrow size distributions, while maximizing the conversion of reactants and enabling recycling of the solvents.

638881 Elsevier Lifetime Achievement Award: Fluidization Centennial -- Reflection on the Past and Prospection in the Future

Abstract:

2021 marks the 100 anniversary of Fluidization and the time is right to reflect on the history and then to prospect into the future. What one can learn from the past and what we should expect in the future.

Since the inception, the "original" gas-solid "conventional" fluidized bed has been expanded, to liquid-solid and three-phase fluidization by changing the fluidizing medium, to circulating fluidization by changing (increasing) the fluidizing velocity beyond terminal velocity, to downflow fluidization by changing the fluid flow direction, and to inverse fluidization by changing the relative density of the fluid. Taking advantages of its large fluid-particle interfacial area, vigorous mixing, and excellent heat and mass transfer, fluidized bed processes have had major successes in the past 100 years, such as fluid catalytic cracking of crude oil, the combustion of coal, and many catalytic chemical reactions.

Recent developments have been pointing to intensifying and integrating fluidization processes, instituting new processes in other "non-conventional" process industries, comprehensive process modelling, and more in-depth fundamental studies to reveal the underlining principles fully utilizing the rapid developments of powerful measurement tools and computational capabilities. Utilizing fine group C particles and establishing micro flow fluidized bed reactors shall also come to play. New fluidization theories, especially those related to the above-mentioned developments, can be anticipated.

638882 PSRI Lectureship Award in Fluidization: CFD as a tool to better understand multiphase flow physics

Abstract:

Science is built on three pillars — experimentation, theory, and computation — which serve as the foundation of scientific discovery. Of these, computation is the fastest growing due to cheap computer hardware and free operating systems and compilers. Much sophisticated software, including computational fluid dynamics (CFD), is low-cost and open source, thus accelerating the use and development of faster, more efficient computational tools. Surprisingly, even with the latest developments in experimentation, theory, and computation, the physics of multiphase flows are still poorly understood. This talk will focus on using CFD to help develop a better understanding of the apparent complexity of fluidized bed operations. The initial part of the talk will focus on analyzing the predictions of different continuum two-fluid methods (TFM) based on some simple examples. Then, we'll switch to discrete methods that provide detailed flow visualizations of particle rotation and interactions with walls to analyze attrition and erosion for high-speed flows. The last example will demonstrate the ability of CFD to provide a clear explanation to the formation and growth of granular jets previously observed experimentally in the past decade. Finally, we will briefly discuss views and opinions on the path forward to simulate large-scale industrial fluidized beds.

627421 Boron-Aluminum Nanoenergetics: Substantial Improvement of Energy Release in Blended Nanomaterials

Abstract:

Boron (B) has attracted considerable attention as a potential high energetic material in propulsion, solid fuels, and explosives due to its exceptionally high gravimetric (58 kJ/g) and volumetric (140 kJ/mL) energy densities. However, the native boron oxide layer on its surface reduces the energy release and slows down the oxidation through a diffusion-controlled process. Reactive metals such as aluminium (Al), when mixed with B have shown to improve the energy release, though by relatively small amounts. We hypothesize that the main effect of blending Al as an additive is to chemically reduce the native boron oxide during combustion leading to two beneficial outcomes: more B is available for oxidation at the expense of less energetic Al, and since the reduction of boron oxide by Al is exothermic, there is the additional energetic benefit of the redox reaction. For such benefit to materialize, one must employ nanoparticles, which carry substantially more oxide than microparticles, and the amount of Al added must match the stoichiometric requirement for the boron oxide present.

Here we present a simple but highly efficient method to extract higher amounts of chemical energy from B using Al than from B alone. We study the effect of Al (~70 nm) addition on the energy release and oxidation efficiency of B (~500 nm). The investigation of Al/B blends with different compositions using the thermochemical analysis is performed to discover the blend with a maximum energy release. Stoichiometric calculations with the help of HRTEM and XPS analyses also indicate the similar optimum composition of Al in B in order to get maximum energy release. This composition provides maximum energy release and a higher degree of oxidation due to the synergistic effect of B oxidation and redox reaction between Al and B_2O_3. HAADF-STEM-EDS analysis on oxidation products of Al/B proves the formation of ternary oxide in the system, which provides porous channels for oxidation of B, thereby increasing its contact with the oxidizer. Similar analysis of the blends demonstrates both qualitatively and quantitatively that Al and B particles stay closer in hydrocarbons, and therefore, they can also benefit propellants as fuel additives.

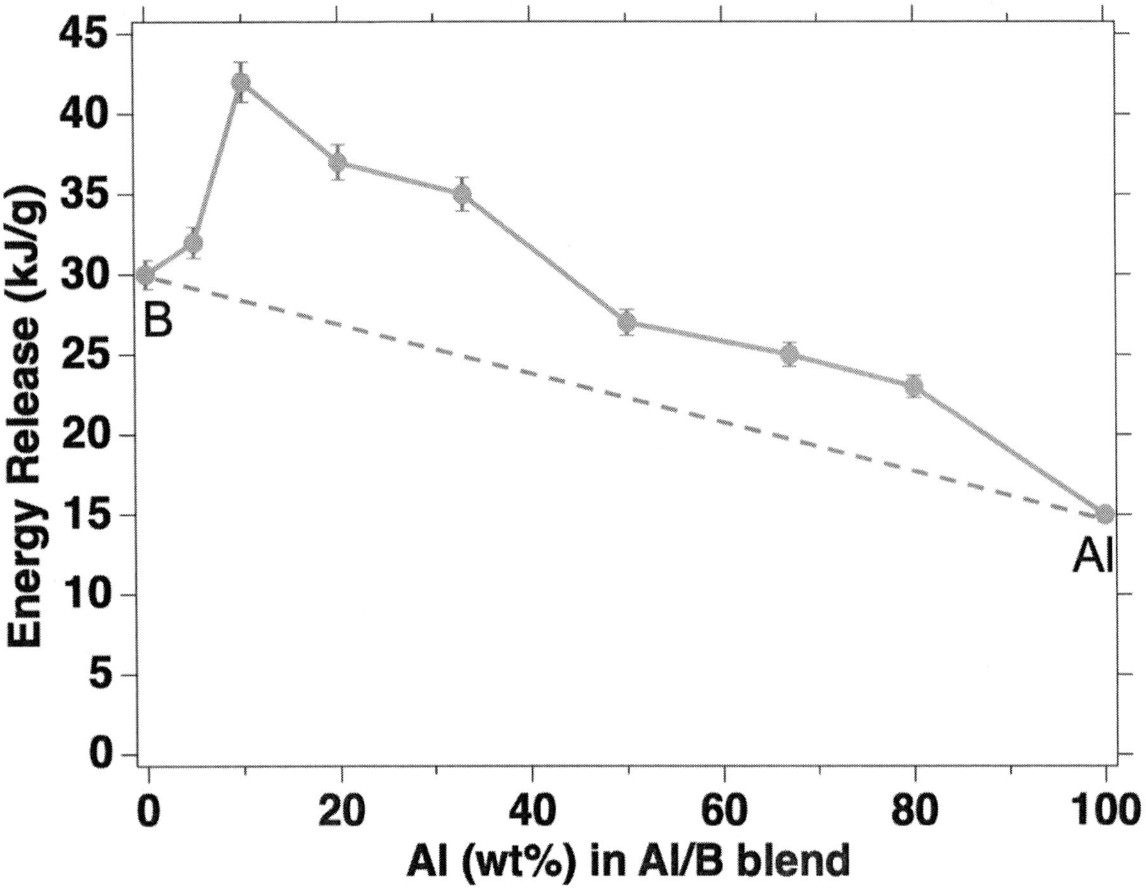
45
40
35
30
25
20
15
10
5
0
Energy Release (kJ/g)
B
Al
0
20
40
60
80
100
Al (wt%) in Al/B blend

629750 Utilizing Meniscus Guided Coating Techniques to Obtain Polymorphic Control and Large Area Single Crystalline Domains in Energetic Material Thin Films

Abstract:

Energetic materials (EMs) are a class of materials that contain a significant amount of stored energy in the chemical bonds within their molecules. This energy is intended to be released upon an initiation event (i.e. impact, friction, shock, spark, heat) for a variety of applications including construction, mining, space exploration, pyrotechnics, and in the military. Undesired initiation can lead to catastrophic and detrimental failures; therefore, these materials require the utmost level of fundamental understanding in terms of predictability and stability. Producing these materials in crystalline form stabilizes the periodic structure of the EM and allows for more consistent material properties (i.e. impact stability, detonation velocity). However, organic crystalline structures can be crystallized in different packing structures, termed polymorphs. Polymorphs exhibit varying material properties due to differences in their crystal structure (i.e. density, morphology); therefore, it is crucial to develop a fundamental understanding of how polymorphs form and how they can interconvert for energetic material stability.

In other fields (i.e. semiconductors, pharmaceuticals), meniscus guided coating has been utilized to obtain polymorphic control in small organic molecule thin films. A specific type of meniscus guided coating, termed solution shearing, sandwiches a solution containing the monomer between a heated substrate and a coating blade that moves translationally at a set speed. Varying the temperature and the coating speed influences evaporation and deposition rates, allowing for the formation of metastable polymorphs. Through solution shearing in a variety of coating regimes, we have crystallized large, aligned, millimeter to centimeter wide crystal domains of two energetic materials of interest. Furthermore, we have utilized Raman spectroscopy to demonstrate polymorphic control for these two EMs. Finally, we have utilized optical microscopy to observe recrystallization of the metastable polymorph of one of the EMs formed *via* solution shearing, and we utilized *in* situ grazing incidence x-ray diffraction (GIXD) to study what type of recrystallization (i.e. polymorphic interconversion, Ostwald's ripening) is observed for thin films of this material.

628154 Application of a Neural Network Approach to Predict the Filtered Drag Force for Gas-Solid Flows

Abstract:

Gas-solid flows with complicated meso-scale inhomogeneities are common in energy and chemical industry. To accurately predict the filtered drag force, filtered two-fluid models have been developed based on correlative analysis of fine-grid two fluid model simulation data. However, these traditional models are found to be unsatisfactory in the prior analysis of predicting the filtered drag force. Recently, data science, especially deep learning, has been used to solve challenge problems in the field of fluid mechanics. In our work, a deep learning approach based on a proposed artificial neural network (ANN) is exploited to predict the filtered drag force for coarse-grid simulations. The proposed ANN model includes an input layer, hidden layers, and an output layer. The hidden layers are divided into three convolutional layers, one flat layer, and two fully-connected layers. The correlation coefficient analyses and probability distribution function of the relative error are used to evaluate the performance of the proposed ANN model. It is found that the proposed ANN model is significantly superior to the multi-layered perceptron neural network model as well as the best available traditional functional model named the dynamic scale-similarity model. The effect of three important parameters, namely, the domain-averaged solid volume fractions, filter size, and convolutional kernel size on the filtered drag force is analysed. The predictability of the proposed ANN model generally increases with the decrease of the domain-averaged solid volume fractions and the increase of the filter size. The predicted results are particularly better when the convolutional kernel size is chosen as three times of the filter size. This finding is significant and indicates that the local filtered drag force is affected profoundly by the flow properties in the near neighborhood in coarse-grid simulations. To better understand the underlying physics behind this finding, a budget analysis of the filtered momentum equation for the solid phase is performed. Moreover, the inclusion of the lateral gas-phase pressure gradient that is perpendicular to the flow direction is also helpful in accurate prediction of the filtered drag force.

Reference

1. Jiang, Y., Kolehmainen, J., Gu, Y., Kevrekidis, Y.G., Ozel, A., Sundaresan, S., "Neural-network-based filtered drag model for gas-particle flows," Powder technology, 346, pp. 403-413 (2019).
2. Ozel, A., Gu, Y., Milioli, C.C., Kolehmainen, J., Sundaresan, S., "Towards filtered drag force model for non-cohesive and cohesive particle-gas flows," Physics of Fluids, 29 (10), pp. 103308 (2017).

632711 Development and Validation of a Comprehensive Drag Model for Simulating Gas-Solid Fluidization in Dense and Dilute Regimes

Abstract:

Two-fluid modeling has been widely used to simulate large-scale fluidized beds, and exhibiting a significant dependence on drag models. In this study, a comprehensive drag model including the whole flow regimes for gas-particle flows is proposed to simulate the hydrodynamics of gas-solid two-phase flow. It combines both the Wen & Yu drag model and the Ergun drag model by a method that is theoretically according to flow regimes. This new drag model is validated with experimental data from spout fluidized bed, bubbling fluidized beds and riser, and compared with simulation results predicted with previous drag models. A computational study was carried out by means of a Two-Fluid model (TFM) based on the kinetic theory of granular flow (KTGF) to explore the influence of the interactions between gas and particles on the formation of heterogeneous flow structure. It is found that the experimental and computed particle velocity profiles show good agreement when using this proposed model in the spout fluidized bed. A better prediction of the solid concentration near the walls of the fluidized bed was obtained due to its favorable feature of dealing with particle concentration transition.

Keywords: Fluidized bed, Two-fluid model, Drag model, Kinetic theory of granular flow

629341 Spatially-Averaged Two-Fluid Models for Heat and Mass Transfer

Abstract:

Clusters are observed to form in gas-particle flows in the presence of a mean body force, such as gravity. Simulations on coarse grids are commonly used to predict the macroscale flow properties of moderately dense gas-particle flows such as full-scale fluidized beds or risers. However, often these grids are too coarse to resolve the mesoscale particle clusters, which can only be a few particle diameters wide [1]. As was shown in many previous studies, the influence of these mesoscale structures on the macroscale flow properties is important, i.e. the bed expansion is severely overestimated if drag models derived for homogeneous suspensions are applied [2]. Therefore, various studies focused on deriving models for the drag reduction. Our approach, for example, is based on spatial averaging, which allows us to model the unresolved contributions in coarse grid simulations [3,4]. However, although various works focused on the drag-force correction, a significantly smaller number discusses heat and mass transfer corrections [5], which are also needed in order to correctly predict the thermodynamics and reactions in coarse-grid simulations.

Previously, we conducted an a priori study on the influence of the particle clusters on the interphase heat transfer [6]. We showed that the resolved transfer can be corrected by a construct similar to the drift velocity [7], which we called drift temperature. This drift temperature is essentially the gas-phase temperature fluctuations as seen by the particles. We expressed it as a function of the variance of the gas-phase temperature and the variance of the solid volume fraction scaled by linear correlation coefficients and derived transport equations for these variances. Furthermore, our a priori analysis showed, that the correlation coefficients can be determined locally and dynamically in coarse grid simulations through the application of test-filters [8].

In a consecutive a posteriori study, we considered different test cases of lab-scale fluidized beds as well as unbound fluidization and validated the previously derived models for momentum and heat transfer. Therefore, we implemented the derived SATFM turbulence models in a modified solver based on the OpenFoam [9] solver twoPhaseEulerFoam [10]. We observed, that the overall heat transfer is overestimated in coarse grid simulations, similar to the overprediction of the drag force. The corrections based on our spatially-averaged approach are in good agreement with the measurements of the interphase heat transfer based on fine-grid simulations of the same cases (see Figure 1).
Furthermore, we expanded this approach to mass transfer, where, as a first step, we again conducted an a priori study and found that a similar drift scalar approach can be used.

References:

[1] Agrawal, K., Loezos, P. N., Syamlal, M., Sundaresan, S., 2001. The role of meso-scale structures in rapid gas-solid flows. J. Fluid Mech. 445, 151–185.
[2] Schneiderbauer, S., Pirker, S., 2014. Filtered and heterogeneity based subgrid modifications for gas-solid drag and solids stresses in bubbling fluidized beds. AIChE J. 60(3), 839–854.
[3] Schneiderbauer, S., 2017. A Spatially-Averaged Two-Fluid Model for Dense Large-Scale Gas-Solid Flows. AIChE J. 63 (8), 3544–3562.
[4] Rauchenzauner S., Schneiderbauer S., 2020. A dynamic Anisotropic Spatially-Averaged Two-Fluid Model for moderately dense gas-particle flows, IJMF 126, 103237.

[5] Sundaresan S., Ozel A., Kolehmainen J., 2018. Toward Constitutive Models for Momentum, Species, and Energy Transport in Gas–Particle Flows, Annu. Rev. Chem. Biomol. Eng. 9 (1), 61–81.
[6] Rauchenzauner S., Schneiderbauer S., 2020. A dynamic Spatially Averaged Two-Fluid Model for heat transport in moderately dense gas–particle flows. Physics of Fluids 32, 063307.
[7] Parmentier, J.-F., Simonin, O., Delsart, O., 2012. A functional subgrid drift velocity model for filtered drag prediction in dense fluidized bed. AIChE J. 58 (4), 1084– 1098.
[8] Lilly, D. K., 1992. A proposed modification of the Germano subgrid-scale closure method. Phys. Fluids A 4 (3), 633–635.
[9] OpenFoam v6 User Guide, The OpenFOAM Foundation, https://cfd.direct/openfoam/user-guide-v6
[10] twoPhaseEulerTurbFoam is available on github,

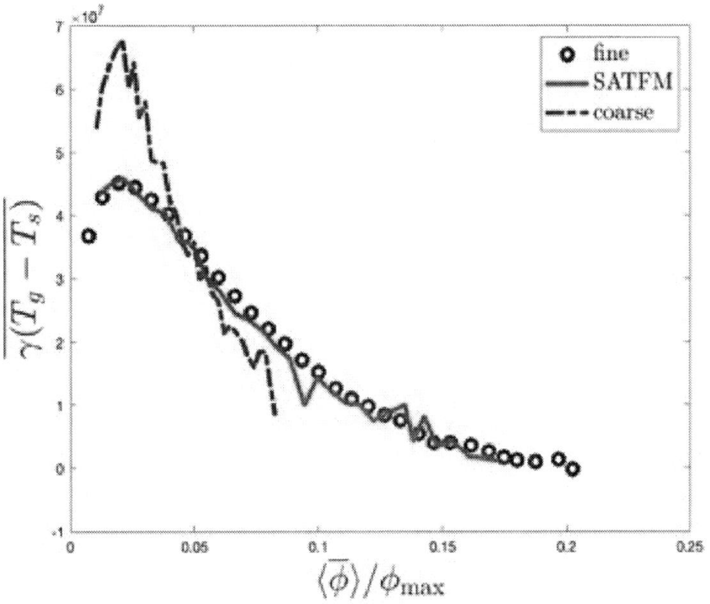

Figure 1: A posteriori evaluation of the interphase heat transfer as a function of the solid volume fraction in the case of unbound fluidization of Geldart Type A particles with an average solid volume fraction of 0.15. The black circles correspond to the measurements of the filtered heat transfer in fine grid simulations, the blue dashed line corresponds to measurements of the filtered heat transfer in coarse grid simulations without corrections and the red line corresponds to measurements of the filtered heat transfer in coarse grid simulations by application of the correction models derived in our SATFM approach.

623429 Enhancement of Fluidization of Fine Powders

Abstract:

Gas-solid fluidization is one of the most broadly used techniques in the industry, where homogenous mixing and vigorous interphase contact are of the primary interest. Fine powders (from nanos to a few microns) of a large surface area are typically good candidates for catalytic materials, but their fluidization is problematic, very hard to achieve. At this scale, interparticle forces, such as capillary force and van der Waals force, play a significant role, making the powders cohesive. As a result, an assembly of fine powders tends to agglomerate and creates gas channels when fluidized at a steady flow, leading to rather poor contacts with the supplied gas [1]. Mitigating agglomeration and breaking up gas channels are, obviously, the key steps to improve their fluidization performance.

Introducing additional degrees of freedom provides flexibility to manipulate fluidization behavior. In the last decade, many works demonstrated the successful uses of external actuators, across different force types and energy sources, to enhance fluidization behavior [2, 3], as shown in Fig. 1. In this work, we experimentally investigate industrially relevant assistance methods that are prevalent, such as vibration, pulsation and stirrer mixing. X-ray computational tomography and pressure probes are used to visualize and characterize the hydrodynamics with particular interests to monitor the effects on agglomeration and gas channels. By pinpointing the degree-of-enhancement induced and uncovering the underlying mechanism, the results will allow us to optimize existing reactors, and design a new class of efficient production process accordingly.

[1] J.R. van Ommen, J. Nijenhuis, M.-O. Coppens, Reshaping the structure of fluidized beds, CEP (2009) 49-57.

[2] V. Francia, K. Wu, M.-O. Coppens, Dynamically structured fluidization: Oscillating the gas flow and other opportunities to intensify gas-solid fluidized bed operation, Chem. Eng. Process. 159 (2021) 108143.

[3] J.R. van Ommen, J.M. Valverde, R. Pfeffer, Fluidization of nanopowders: a review, J. Nanopart. Res. 14 (2012) 737.

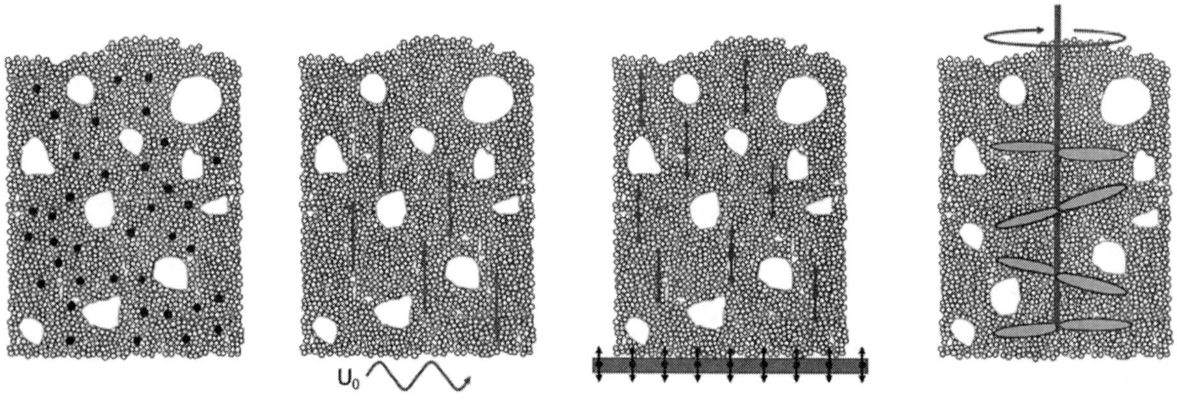

Figure 1 – Schematic of methods to assist the fluidization of fine powders. Adapted based on [2].

623439 Pressure Fluctuations in Gas-Solid Fluidized Beds: Measurement and Analysis

Abstract:

Pressure measurements are widely applied in commercial gas-solid fluidized beds for process supervision and control. Although tomography techniques were developed for medical research, such as magnetic resonance imaging (MRI), electrical capacitance volume tomography (ECVT) and X-Ray computed tomography (XRT), chemical engineering have successfully applied them to examine the hydrodynamics of gas-solid suspensions. However, they are rather costly, arduous to operate, and, more importantly, impractical at the industrial scale. In contrast, measuring pressure routine, robust, flexible, and inexpensive. Nevertheless, interpreting pressure fluctuation signals is non-trivial. Mapping measured signals to the corresponding hydrodynamic events requires a prior knowledge, and together with a proper selection of the methods of analysis to decompose the measured signals and extract essential characteristics.

In this contribution, we review the measurements of pressure fluctuations in gas-solid fluidized beds, concerning the cause of the fluctuations [1], mathematical methods and recent applications, including the studies in the domain of time, frequency and state, as shown in Fig. 1 [2, 3]. We also demonstrate the typical source of errors and the special cares to consider, intending to provide a systematic yet straightforward guide for carrying out pressure measurements in gas-solid fluidized beds.

[1] J. van der Schaaf, J. Schouten, C. van den Bleek, Origin, propagation and attenuation of pressure waves in gas—solid fluidized beds, Powder Technol. 95 (1998) 220-233.

[2] J.R. van Ommen, S. Sasic, J. van der Schaaf, S. Gheorghiu, F. Johnsson, M.-O. Coppens, Time-series analysis of pressure fluctuations in gas–solid fluidized beds – A review, Int. J. Multiphase Flow 37 (2011) 403-428.

[3] F. Johnsson, R.C. Zijerveld, J.C. Schouten, C.M. van den Bleek, B. Leckner, Characterization of fluidization regimes by time-series analysis of pressure fluctuations, Int. J. Multiphase Flow 26 (2000) 663-715.

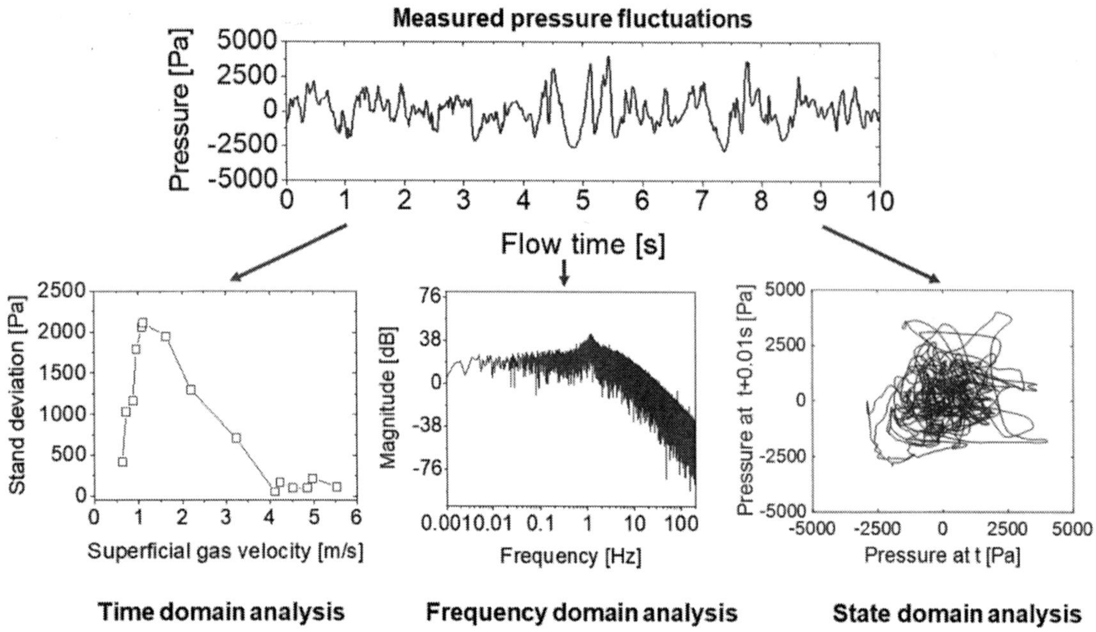

Figure 1 – Typical methods to analyze a time-series of pressure fluctuation in a gas-solid bubbling fluidized bed. The pressure flucuations data are taken from [3].

629252 Fluidization of Wet Granular Materials

Abstract:

Fine and ultrafine powders have recently received growing interest in both industrial and academic sectors [1, 2] due to their very distinctive features, mainly coming from their very small primary particle size and very large surface-to-volume ratio. Indeed, due to these characteristics, they can provide better contact efficiency and higher reaction rates per unit volume of reactor than traditional materials in the case of gas/solid and solid/solid reactions. They have been used to produce a large variety of materials, such as catalysts, sorbents, cosmetics, etc. Therefore, the interest in using this type of granular materials in a variety of industrial processes raises many questions about how they can be handled and processed in large scale applications. Among many available techniques for continuously handling and dispersing granular solids, gas fluidization is one of the most efficient one used in industries, mainly due to the large gas–solid contact area.

Relatively small changes in saturation of interstitial liquid can transform a fine free-flowing powder into one which is cohesive. The liquid capillary bridges formed connecting grains can alter the physical properties of the granular materials dramatically [3, 4]. The attractive forces between the particles connected by bridges in the micro scale results in increasing bulk cohesion for moderately saturated materials. As the amount of liquid increases, the capillary bridges start to merge, and they eventually disappear altogether when the powder is fully saturated. A typical fluidization curve for dry powders exhibits fixed bed characterized by linear increase in pressure drop with increasing air flow velocity. This is followed by a fluidized bed with a constant pressure drop with further increase in air velocity. However, wet cohesive powders are likely to display different fluidization behaviour due to the inhomogeneous gas flow and bubbling [5]. The fluidization behaviour of the powder upon addition of small amounts of silicone oil liquid has been tested for different grain sizes. The aim of our work is to distinguish different regimes of fluidizations exhibited by cohesive powders for different saturation of silicone oil and for powders of different sizes. We compare dry non-cohesive powder fluidization behavior with wet moderately-to-strongly saturated cohesive glass powders.

A virtual Couette cell is used to study the behaviour of fluidization of powders of varying cohesion intensity. This rheometer was originally developed by Ait Ali Yahia et al. [6] to study the fluidization behaviour of dry non-cohesive glass powders. We generalize the use of this Couette cell rheometer for experimenting fluidization behaviour of both dry non-cohesive and wet cohesive materials.

[1] Zhu, C, Q Yu, RN Dave, R Pfeffer. Gas fluidization characteristics of nanoparticle agglomerates. AIChE Journal. 2005; 51(2): 426-439.[2] van Ommen, JR, JM Valverde, R Pfeffer. Fluidization of nanopowders: a review. Journal of Nanoparticle Research. 2012; 14(3): 1-29.[3] Herminghaus, S. Dynamics of wet granular materials. Advances in Physics. 2005; 54(3): 221-261.[4] Roy, S, A Singh, S Luding, T Weinhart. Micro-macro transition and simplified contact models for wet granular materials. Computational Particle Mechanics. 2016; 3(4): 449-462.[5] Wang, YF, QG Cheng, KH Zhang, YQ Zhong, ZX Luo. Study of fluidized characteristics of rock avalanches under effect of entrapped air. Rock and Soil Mechanics. 2014; 35(10): 2775-2786.[6] Ait Ali Yahia, L, TM Piepke, B Ross, A Ozel, R Ocone. Development of a virtual Couette rheometer for aerated granular material. AIChE Journal. 2020; 66(6): e16945.

621559 Effect of Scale up Process on Reactor Performance within Riser: Simulation Using Ozone Decomposition

Abstract:

Introduction

Circulating fluidized beds (CFB) are widely used in chemical and energy fields because of the high mass and heat transfer rate, significantly reduced back mixing. Successful commercialization of CFBs relies on the ability to scale up the reactors from lab to industrial scale especially for the risers. To apply fundamental knowledge from pilot scale risers into a commercial riser, it is necessary to achieve a clear understanding of the scale-up effects. The scale up effect on the hydrodynamics within riser reactors has been investigated in many aspects, such as solids concentration distribution, particle velocity, solids flux distribution, flow development and regime transition. On the other hand, scaling laws are put forward in order to ensure the hydrodynamic similarity within riser during the scale up process. The scaling laws are deduced by the dimensional analysis of the momentum balance on a particle. The cluster features would change with the variation of particle properties during scaling process and have an influence on the hydrodynamics similarity. In general, recently researches about riser scale up process are focus on the aspect of hydrodynamics. However, the chemical similarity in terms of conversion and selectivity also has to be considered during the scale up process of chemical reactor. The chemical conversion and selectivity depend on the interactions of mass transfer, kinetics, and hydrodynamics, the research results of bubbling fluidized bed have shown that a scale change will always directly influence the ratio of these interactions. Therefore, the effect of scale up process on reactor performance within riser should be investigated.

The reactor performance within riser is always evaluated by the detection of ozone decomposition. Researchers have investigated the local reactor performance and the effect of operation condition, internal structures, new type structures on the reactor performance. The axial and lateral ozone concentration distribution are the most common used to analyze reactor performance.

CFD simulation has shown to be an effective tool in the evaluation of scaling rules with the consideration of cluster effect for risers, since the simulation method offers the advantage by facilitating evaluation of these scaling parameters with fewer requirements of extensive physical experiments. Meanwhile, the ozone decomposition process is also could be realized by the CFD simulation method. Therefore, the effect of scale up process on the reactor performance is investigated by the simulation of ozone decomposition process in the scaling risers with the guide of scaling rules.

Glicksman's full and simplified sets of scaling parameters are used to build two 1:3 scaled-up risers, which is the same as the previous research. The energy minimization multiscale (EMMS) drag model is used to describe the gas solids momentum interaction, since it can capture the effect of clusters on the gas solids interaction. The EMMS drag force coefficient for each case coupled with the two-fluid models and kinetic temperature granular flow model (KTGF) is used to simulate the flow structures in two-dimension risers. The ozone decomposition was realized within the risers through the coupling with a user defined function (UDF) which describe the ozone decomposition reaction rate equation. The effect of scale up process on the reactor performance especially for the gas solids contact efficiency was analyzed by the comparison of dimensionless axial and radial ozone concentration distribution and corresponding solids concentration profiles between two scales. Furthermore, the cluster characteristics in each case were used to explain the variation of reactor performance during scaling up process.

Conclusions

Ozone decomposition in the riser is simulated with CFD in order to explore the scale up effect on the chemical reaction performance. The variation of gas solids contact efficiency is revealed by the comparison of cluster effect in each set up. The ozone decomposition is effect by the solids volume fraction and gas phase residence time at the same time. The ozone decomposition degree in the scaling riser with full set scaling rules is higher than the small riser, while the ozone decomposition degree in the scaling riser with simplified rules is lower than the small riser. The radial ozone uniformity becomes worse during the scale up process. The overall gas solids contact efficiency is improved during the scale up process with full set scaling rule. The effect of scale up on the gas solids contact efficiency in the core region is not obvious, while the effect of scale up on the gas solids contact efficiency in the wall region is related to the variation of cluster effect during scale up process.

624110 X-Ray Computed Tomography-Based Micro-Porosity Analysis and Discrete Element Modeling: Algorithm Development and Application for Woody Biomass

Abstract:

The properties of lignocellulosic biomass, such as bulk physical, thermal, and mechanical properties, as well as the mobility of enzymes or catalysts, are largely affected by porosity, pore structures, and pore size distribution. While X-ray computed tomography (CT) has been introduced to effectively produce 3D volumetric images of material microstructure, a quantitative porosity analysis from the 3D images has never been done. This work presents a first-of-its-kind X-ray CT-based quantitative micro-porosity analysis method and applies the developed toolkit to characterize the microstructure and internal porosity distribution of biomass particles. Building on the microstructure data, a high-fidelity micromechanical discrete element model is developed to simulate the mechanical behavior of the particle. In the application for woody biomass, sample loblolly pine chips are scanned by a nano-resolution X-ray CT system and digitally reconstructed after a sequence of image processing operations. A comprehensive porosity analysis is then performed to quantify the envelope porosity, the spatial distribution of local porosity, and the directional porosity. Then, micromechanical simulations are performed using the developed discrete element model to predict the mechanical and fracturing behavior of loblolly pine particles. Solutions to various challenges in image processing, porosity calculation, large data handling, and microstructure-based mechanical model development are provided.

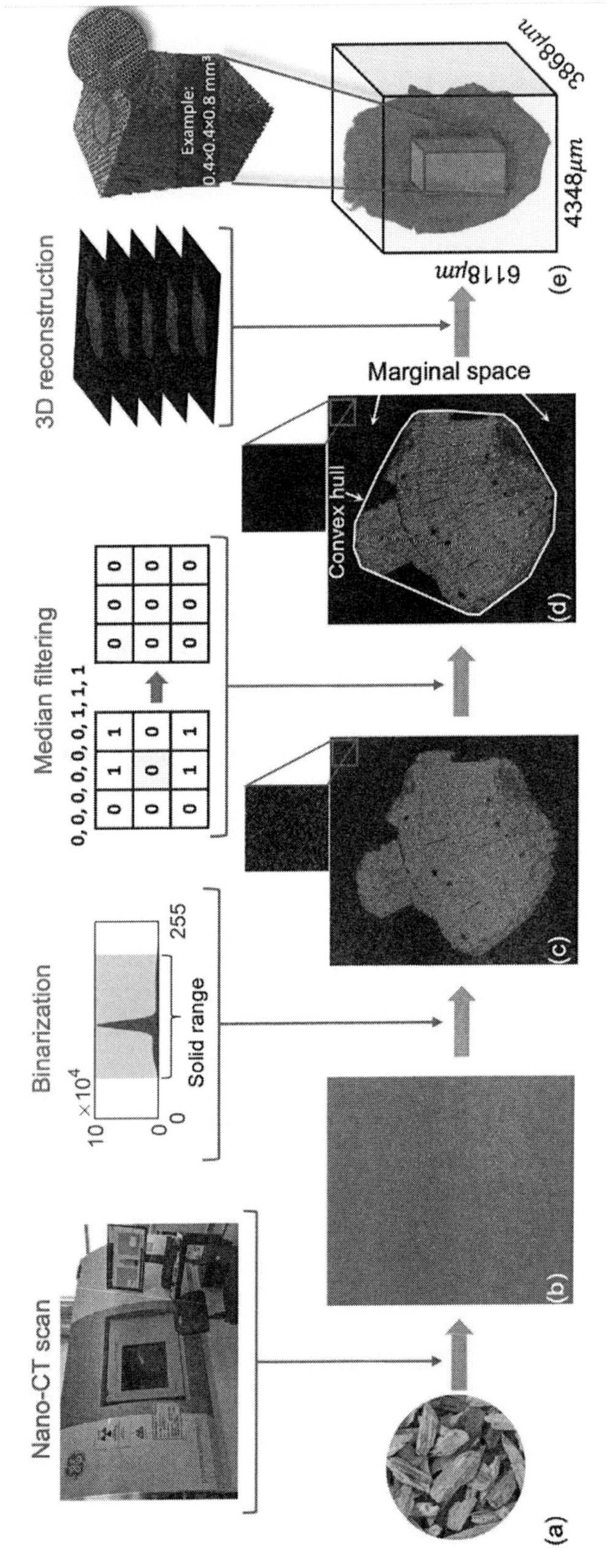

Nano-CT scan
Binarization
Median filtering
3D reconstruction
×10^4
10
0
0
255
Solid range
0, 0, 0, 0, 0, 0, 1, 1, 1
Convex hull
Marginal space
Example:
0.4×0.4×0.8 mm^3
6118μm
4348μm
3868μm
(a)
(b)
(c)
(d)
(e)

628330 Continuum Simulations of Granular Flow Near the Maximum Packing Limit Using a Novel Solution Approach to Address Realizability

Abstract:

The hydrodynamics and rheology of powder flow are critical in many industrial processes such as pharmaceutical and agricultural production, and petrochemical refining. Continuum simulations solve the averaged equations for conservation of mass and momentum by treating the granular medium as a continuum to predict granular hydrodynamics. A comparison of continuum simulations of silo discharge (Vidyapati *et al.* 2013) with experimental data and discrete element model simulations indicates that the simulations are limited in their accuracy of predicted discharge rates by the constitutive model for granular stress. One of the challenges in incorporating newly proposed constitutive models for the granular stress in continuum simulations, which account for intermediate regime behavior (Chialvo *et al.* 2012, Sun *et al.* 2011, Vidyapati *et al.* 2016), is ensuring their realizable behavior near the maximum packing limit. Realizability in this context means guaranteeing that the solid volume fraction is always non-negative and does not exceed its theoretical maximum value without *ad hoc* modifications.

The behavior of continuum solvers near the packing limit depends on the constitutive model for the granular pressure as a function of solid volume fraction. As the solid volume fraction approaches the maximum packing limit, there is a threshold above which the granular material behaves like an incompressible fluid, resulting in high granular pressures. High pressures generate a moving shock that requires special numerical treatment including (1) formulation of the governing equations in terms of conserved variables, and (2) a flux-splitting scheme which is monotone and total-variation-diminishing (TVD) in order to capture the propagating shock waves. This solution method gives accurate results without any *ad hoc* modifications, even when the pressure versus volume fraction curve does not have a steep slope as the volume fraction approaches the maximum packing limit.

The newly developed computational scheme is implemented in OpenFOAM. We report solutions to a three-dimensional (3-D) granular flow problem illustrating the transition from variable density to incompressible regime at the maximum packing limit by extending the previously reported 1-D canonical problem when a sphere consisting of granular particles is compressed isotropically. We verify the solution method using the well-known Sedov Point Blast problem to establish accuracy. Continuum simulations of silo discharge using the novel numerical approach have been compared to experimental data and discrete element model simulations to validate the results, and good agreement is observed.

628359 Optimizing Superparamagnetic Iron Oxide Nanoparticle Synthesis and Peg Coating for Magnetic Particle Imaging Performance and Long Blood Circulation Half Life

Abstract:

Magnetic particle imaging (MPI) is a novel biomedical imaging modality that allows non-invasive, tomographic, and quantitative tracking of the distribution of superparamagnetic iron oxide nanoparticle (SPION) tracers through their nonlinear dynamic magnetization response to an alternating excitation field. As MPI directly detects SPIONs, the resolution and sensitivity of MPI is primarily impacted by the magnetic response of the nanoparticle tracers. Importantly, the physics of signal generation in MPI are distinct from that in Magnetic Resonance Imaging (MRI). As such, SPION tracers developed for MRI are not necessarily ideal for MPI. In addition, surface modification and formulation are also important factors that affect application performance. For example, there are several applications that would benefit from long blood circulation time, such as in blood pool imaging, evaluating traumatic brain injury, gut bleed detection, and functional MPI. These considerations suggest a need for developing SPIONs with physicochemical and magnetic properties that are tailored for specific MPI applications.

Ferucarbotran (Resovist®), a commercially available SPION contrast agent developed specifically for MRI is commonly used for MPI studies. It has been suggested that only 3% of the total iron mass in these tracers contributes to their MPI signal. Another commercially available tracer is Synomag-D®, an aqueous suspension of multi-core maghemite nanoparticles covered by a dextran shell with PEG-OMe on the surface. The MPI performance of Synomag-D® has been evaluated using a magnetic particle spectrometer and results suggest better performance compared to Resovist®. However, studies evaluating its *in vivo* performance are lacking. LS-008, a tailored tracer developed for MPI with high resolution was evaluated to have blood half-life of 105 min in mice. However, LS-008 tracers are no longer available and suitable alternatives that combine resolution with long circulation time are lacking. Furthermore, while these tracers are widely tested using prototype MPI scanners and with the Bruker pre-clinical MPI scanner, their performance has not been evaluated in the newer MOMENTUM™ pre-clinical scanner. Because SPION performance varies with the different scanner construction parameters, such as the amplitude and frequency of the applied magnetic field and the magnitude of the field gradient, the MPI performance of a given tracer can vary from different type of scanners. The growing adoption of the MOMENTUM™ MPI scanner suggests that comparative performance studies of MPI tracers using this scanner would be of value to the community.

Here, we will discuss the physics of signal generation in MPI and how they inform optimization of SPION synthesis for sensitivity and resolution in MPI. We will also discuss the requirements on particle surface coating and colloidal stability imposed by applications of MPI to blood pool imaging. Experimental results will be presented for single-core SPION MPI tracers obtained using a semi-batch thermal decomposition synthesis with controlled addition of molecular oxygen, followed by an optimized PEG-silane ligand exchange procedure. Results will show how MPI performance relates to thermal decomposition synthesis conditions and intrinsic magnetic properties of the SPIONs. The physical and magnetic properties, MPI performance, and blood circulation time of these tracers will be compared to those of the two commercially available SPIONs: ferucarbotran and PEG-coated Synomag®-D. The optimized tracers have better MPI sensitivity and much longer circulation half-life than both commercial tracers, making them ideal for use as blood pool imaging tracer in MPI.

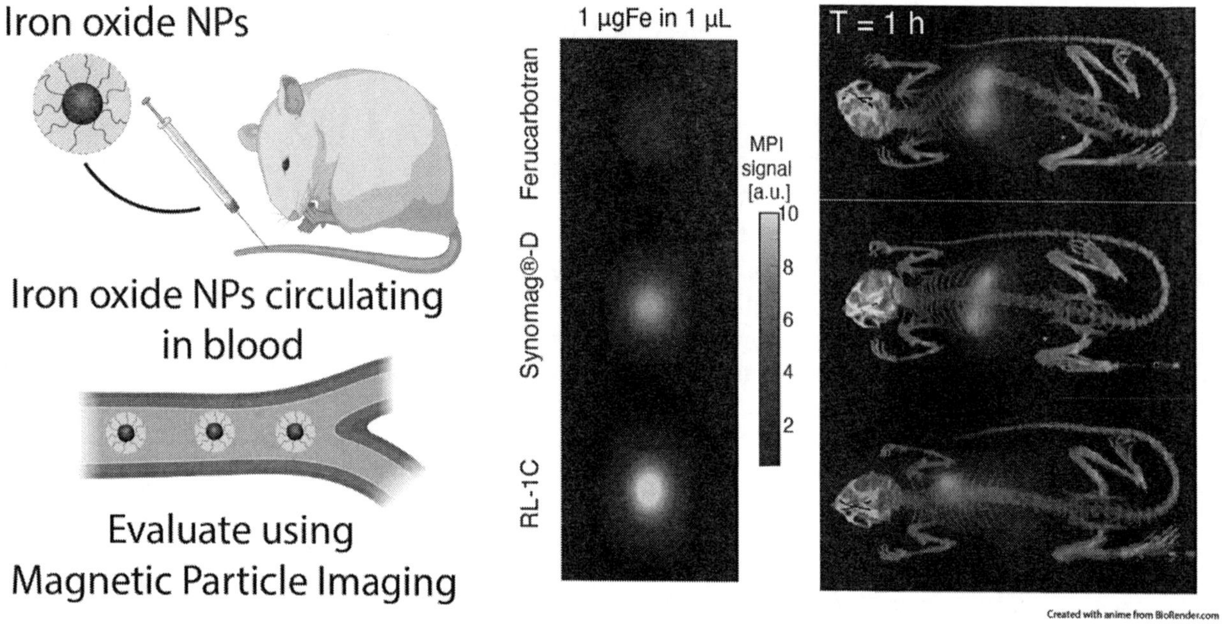
Iron oxide NPs
Iron oxide NPs circulating in blood
Evaluate using Magnetic Particle Imaging
1 μgFe in 1 μL
Ferucarbotran
Synomag®-D
RL-1C
MPI signal [a.u.]
10
8
6
4
2
T = 1 h
Created with anime from BioRender.com

633306 On the Relative Effect of Nanorod Versus Oval-Shaped ZnO Nanoparticles in Their Photocatalytic Activity

Abstract:

Pharmaceutical industries generate waste, that contain a large number of contaminants, which are released into the atmosphere/environment after partial treatments. Another type of waste is municipal waste, where large amounts of antibiotics and other solvents are released because of their low removal efficiency. Generally, these wastes include different drugs that are used in the treatment of various human diseases. The presence of these contaminants in the environment can cause direct effect on human health and other living bodies. Their long-term presence in the open atmosphere can create drug resistance among the microorganisms and as a result, such drugs may not work in a particular health issues. Various methods of removal/degradation have been reported to minimize these contaminants below the permissible limits. Methods such as advanced oxidation processes, catalytic ozonization, photocatalytic degradation, UV-H_2O_2, are majorly reported. Photocatalytic degradation was found to be promising, due to its ease of use, handling a large number of wastes, and being a stable method. In the past, various types of nanoparticle-based photocatalysts such as, TiO_2, ZnO, ZnS, $ZnIn_2S_4$ and gold were reported. Among these, TiO_2 and ZnO are frequently investigated. Their efficiencies can be tuned/tested by their size, shape and surface interactions.

In the present work, ZnO nanoparticles have been used as a photocatalyst, which has attracted researchers, due to their unique physical and chemical characteristics. ZnO has a wide bandgap (E_g = 3.37 eV), high photosensitivity, higher electron mobility, as well as has high quantum efficiency. In addition to that, biocompatibility and eco-friendliness of ZnO, makes it an ideal candidate for the photocatalytic degradation process. Here we have used nanorod (NR) and oval-shaped (OS) ZnO as a photocatalyst, for the degradation of rifampicin (RIF). RIF is a pharmaceutical drug that is mostly used in the treatment of tuberculosis as a first-line drug. These drugs are recalcitrant and therefore their removal is a necessary task for the safety of humans.

For the NR synthesis, equimolar concentration (10 mM) of each of $ZnCl_2$ precursor and hexamethylenetetramine (HMTA), as a reducing agent, were mixed and stirred at 85 °C, for 5 hours. Similarly, for the OS synthesis, the 10 mM $ZnCl_2$ precursor and 40 mM ammonia solution (25 % w/v), as reducing agent, were mixed and stirred at 85 °C for 1 hour. ZnO was formed due to Zn^{2+} ion from the precursor and OH^{-} ion from the reducing agent and at a particular temperature leading the ZnO formation [eq. 1]. X-ray diffraction analysis confirms the hexagonal wurtzite, crystalline ZnO structure. In the case where NH_3 was used as a reducing agent, within a short time (10 second), all the free Zn^{2+} ions were utilized and resulted in ZnO oval-shaped nanoparticles, whereas slow degradation of HMTA lead to formation of ZnO nanorods. Due to this reason, the reaction time is different in both cases. Further, these properties were examined in detail using different characterizing tools. The sizes of NR are found to be- 2190 nm and 367 nm in length and diameter, respectively. while for OS, it is found- 314 nm and 160 nm respectively.

$Zn^{2+} + 2OH^{-} \text{-->} Zn(OH)_2 \text{-->} ZnO_{(s)} + H_2O$ Eq. 1

To determine the efficacy of the ZnO catalyst, we have used both NR and OS shaped nanoparticles, in the photocatalyst degradation of the drug. Our results show that, the degradation rate in NR is high compared to OS. In case of NR, the degradation rate is (0.04 hr-1 mg-1), whereas in OS the rate is only (0.013 hr^{-1} mg^{-1}). The reason for obtaining a higher degradation rate in NR can be due to its higher surface area, lower energy E_g, and higher concentration of oxygen vacancies. The primary reason for the difference in the degradation rate is the BET surface area (SA), in which NR (67.9 m^2 g^{-1}) has 25 times higher area than OS (2.7 m^2 g^{-1}). Also, the BET pore volume of NR is six-fold higher than OS (NR - 0.06 cm^3 g^{-1}; OS - 0.01 cm^3 g^{-1}). Hence, NR has a large number of active sites on its surface, by which easily dissociable constituents were easily degraded. Using DRS, we found that the E_g of NR (3.21 eV) is lower than OS (3.24 eV), so the electronic transitions will use less energy for excitation. With XPS, we found that, the concentration of oxygen vacancies (V_o) is higher in NR compared to OS, and these V_o act as a donor. The electrons are trapped by these donors and due to this, trapping decreases the charge recombination and thereby enhancing the charge separation efficiency in NR. OS shows higher zinc interstitial (Zn_i) defects than NR, but these defects were insignificant in enhancing degradation efficiency.

The valence band (VB) and conduction band (CB) potentials of the photocatalysts NR and OS were calculated theoretically using the Mulliken electronegativity parameter. To calculate it, the standard redox potentials of 2.38 eV (OH^*/H_2O) and -0.33 eV (O_2/O_2^*) were used, as VB and CB positions respectively. We have observed that the VB of both NR and OS is 2.86 and 2.87 eV respectively and is more positive than the standard redox potential. Therefore, the degradation in both could be occurring due to the oxidation of H_2O to OH^* radicals (Fig. 1). The CB potential of NR and OS is -0.35 eV and -0.36 respectively, lower than standard potential. Therefore, electron in CB of both can reduce O_2 and generate the O_2^* radicals. The generated radicals, OH^* and O_2^* are taking part in RIF degradations (Fig. 1). Overall, the higher SA, high V_o, and lower E_g in NR compared to OS, leads to the higher photocatalytic activity, while rest of the parameters are less effective.

Hence, the effect of physicochemical and optical properties of photocatalysts that lead to enhancement of the photocatalytic activity of the drug, RIF, has been investigated.

Figure 1: Photo-degradation of RIF with NR versus OS

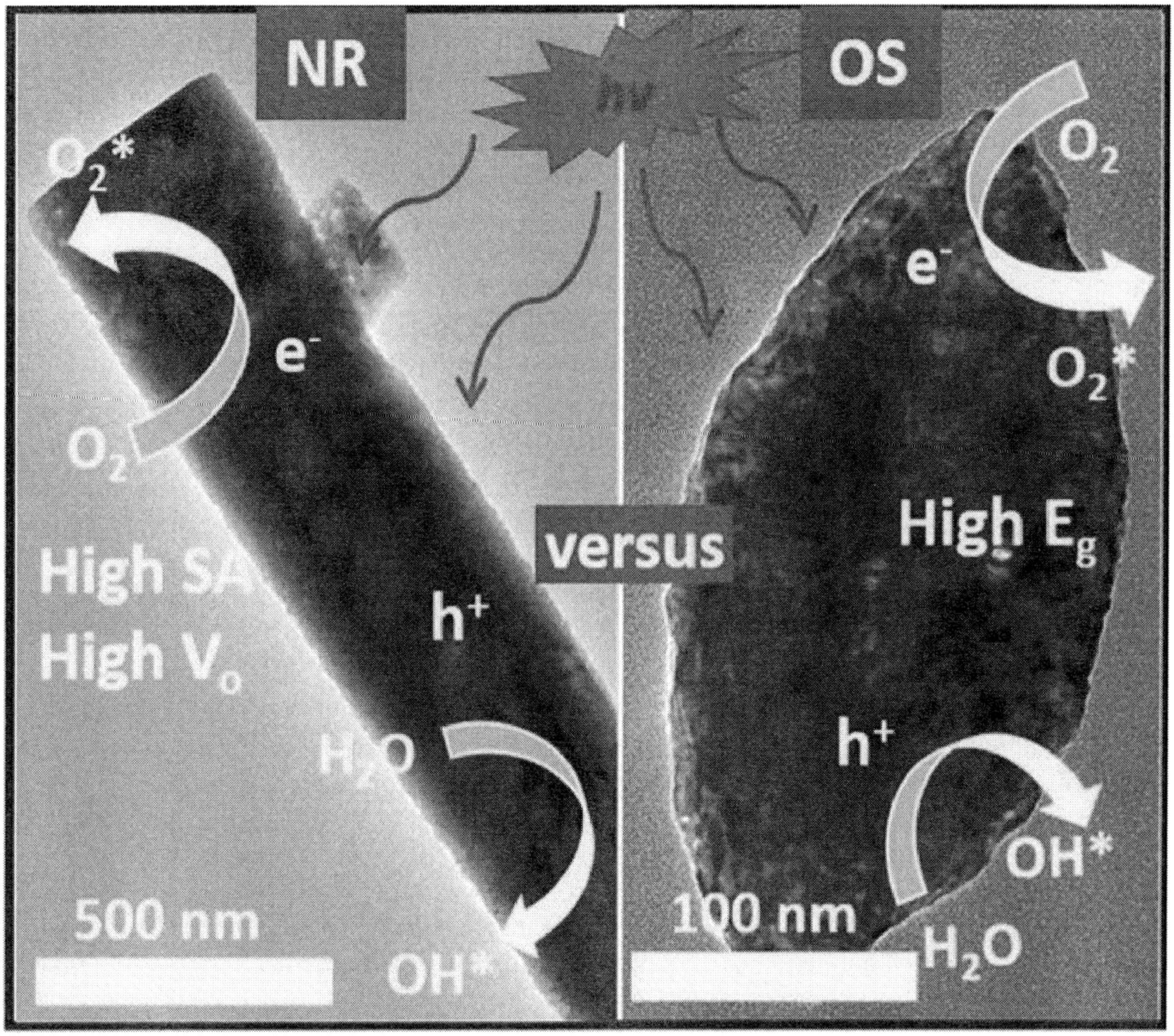
NR
OS
hν
O_2*
O_2
e⁻
O_2
e⁻
O_2*
versus
High E_g
High SA
High V_o
h+
h+
H_2O
OH*
500 nm
100 nm
OH*
H_2O

631876 Biomineralization of Functional AIS/ZnS Quantum Dots

Abstract:

Semiconductor nanocrystals or quantum dots (QDs) are used for a wide range of applications spanning bioimaging, photocatalysis, LEDs, and solar cells due to their unique optical and electronic properties.[1] While II-VI QDs are popular because of their luminescent properties such as high quantum yield, tunable emission wavelength, and narrow emission spectra, their intrinsic toxicities limit their direct usage in optoelectronics and biological systems.[2] Alternative, heavy metal free (i.e., green) I-III-VI chalcopyrite multicomponent semiconductors such as $AgInS_2$ and $CuInS_2$ have attracted considerable interest as a result of their long PL lifetime and tunable composition, the latter enabling tailored spectroscopic properties spanning the UV to the NIR region.[2,3] Among I-III-VI materials, Ag-In-S (AIS) QDs gain considerable attention due to their non-toxic nature, attractive bandgap (1.87 eV) and Bohr diameter (5.5 nm), and the fact that their intrinsic PL quantum efficiencies (~10%) can be tuned to the levels commensurate with II-VI type QDs by the introduction of wide band gap materials (ZnS) as well as size/composition control.[2,3] The synthesis of such multicomponent QDs generally employs hydrothermal, solvothermal and single precursor thermal decomposition processes.[3] The associated high temperatures and/or toxic solvents, however, undercut the overall sustainability of these 'green' materials. Low temperature, aqueous phase biomineralization-based routes to size controlled QDs offer an alternative, 'green', and potentially scalable[4] synthesis strategy.[5]

In this study, we have investigated the sequential synthesis of functional AIS/ZnS QDs with tunable optical properties from biomineralized In-S nanocrystals. Specifically, we demonstrate how aqueous-phase, low-temperature enzymatic (*cystathionine gamma lyase*) turnover of the amino acid L-cysteine yields sufficient endogenous H_2S in buffered solutions of indium chloride for the nucleation of narrowly distributed In-S nanocrystals. Cysteine simultaneously serves as a source of H_2S and stabilizer of the biomineralized In-S throughout the subsequent Ag nitrate-based cation exchange process leading to AIS QDs. We demonstrate how the solution-phase metal precursor ratio (In:Ag) offers a facile knob for directly tuning the composition of the AIS product, and, thereby, the photoluminescence quantum yield (PL QY). We further elucidate and exploit the kinetics of Ag exchange, through controlled incubation, for improving PL lifetime and PL QY, likely by structural annealing. The addition of ZnS more than doubles this PL QY while retaining the optical properties of the AIS core. The resulting AIS/ZnS QDs offer promise for a range of applications. Specifically, we demonstrate their efficacy for visible light photocatalytic H_2 generation. In addition, we confirm the non-toxic nature of the AIS/ZnS QDs through cell viability studies and, upon decoration with key antibodies, show their suitability for targeted bioimaging of THP-1 macrophages. Ultimately, this study advances the versatility of single enzyme biomineralization as a means for the more broadly sustainable and potentially scalable synthesis of various 'green', non-toxic metal sulfide semiconductors for applications spanning photocatalysis to bioimaging.

629117 CG-MD and CFD Simulation of Continuous Manufacturing of Liposome Formation

Abstract:

Purpose:

Liposomes is one of the most researched nanocarriers for drug delivery applications and have major benefits such as stabilizing therapeutic compounds, enhancing drug uptake, and improving biodistribution to target sites. In brief, liposomes are spherical vesicles comprising amphipathic phospholipids either as a single bilayer or as a concentric series of multiple bilayers surrounding an aqueous core. Drug molecules can be encapsulated in the aqueous core and/or loaded into the lipid bilayer. Conventionally, liposomes are processed and manufactured by batch processing in the pharmaceutical industry. However, some disadvantages of batch processing such as scalability and reproducibility issues have led to alternative approaches. Continuous liposome processing is one approach that provides many benefits, such as higher-throughput, increased productivity and reduced energy requirements. An innovative continuous liposome processing system was developed at UConn, where two liquid flows are mixed in a highly controlled manner *via* a co-axial turbulent jet, producing monodispersed nanoparticles. Understanding the mixing process and lipid aggregation and liposomal formation on a molecular level will provide better insight to macroscopic attributes such as particle size and liposomal stability. In this work, we studied the effect of intermolecular forces (e.g., dipole-dipole forces) among the various components (ethanol, lipid and water) along with applying the principles of fluid dynamics. Accordingly, we applied a multi-scale computational study of the liposome formation process as a coaxial turbulent jet flow to probe the underlying mechanism and to quantitatively predict the formation of liposomal nanoparticles.

Methods:

Both coarse-grained molecular dynamics (CG-MD), as a micro-scale investigation, and computational fluid dynamics (CFD), as a meso-scale simulation, have been conducted to reveal the detailed mechanism of liposome formation, and also implement multi-scale case studies for the process. The MD simulation trajectories and their analysis were carried out using GROMACS package. We applied MARTINI force-fields for CG-MD simulations along with periodic boundary conditions. In addition, the particle size changes were studied in Non-Equilibrium Molecular Dynamics (NEMD) simulations with different flow shear rates. Large Eddy Simulation (LES) model in COMSOL Multiphysics incorporating energy equation, in CFD simulations, were implemented with a high-resolution mesh in the mixing area. The simulation results were verified with experiments, using flow patterns and temperature profiles.

Results:

From the CG-MD simulations, liposomes can be formed successfully in co-solvent solution. Though the MARTINI force fields (FFs) could not capture the experimental particle size, the liposome formation mechanism can be revealed at different ethanol percentages and different process temperatures. Additionally, the bilayer thickness of formed micelles at different temperatures agrees well with the trend of thickness changes from previous experimental studies. From NEMD simulations, the ratio of particle size changes caused by shear stress matched with previous experimental results.

By refining the CFD element sizes, a mesh independence study was done to manifest the CFD solution's independence of the mesh resolution. Moreover, an increased formation of vortices led to smaller particles being formed at higher flow rates has been shown from vortex identification studies and probability density functions. Therefore, the multi-scale simulations were found to be accurate when compared to the experimental data and trends.

Conclusions:

Although, it was a challenge to model the formation of large sized liposomes, the results from CG-MD accurately portray liposome internal structure view, bilayer thickness, and formation mechanisms, which are difficult to observe and study through experimental work. Furthermore, the impact of process temperatures and flow rates to the jet flow have been captured by CFD simulations. The modeled MD shear flow indicates the breakdown of particles are consistent with eddies generation in CFD simulations, and further the ratio of particle size changes, in CG-MD, agrees with the measurements from experimental work. The integrative discrete CG-MD and continuum CFD computational modeling of liposome formation were found to be two viable complementary approaches which cover micro-scale and meso-scale, respectively.

Keywords:

Continuous processing, Liposomes, Molecular dynamics, Multiscale, Computational fluid dynamics

Acknowledgements: FDA Grant# 1U01FD005773-01.

Disclaimer: This article reflects the views of the authors and should not be construed to represent FDA's views or policies.

630380 Facet Effects Define Structure-Activity Relationships of Ultra-Low Loading Pt/CeO_2 Catalysts with Atomic Dispersion

Abstract:

A well-defined surface structure is key to exploring structure-activity relationships in heterogeneous catalysis. By combining single atom deposition and metal oxide nanoshape supports, well-defined Pt/CeO_2 catalysts with ultra-low loading were prepared using a modified Atomic Layer Deposition (ALD) technique to investigate surface facet effects of ceria on single atoms Pt under reduction conditions. Three CeO_2 shape supports, including octahedra, rods, cubes, were synthesized by a surfactant-free hydrothermal method [1]. The benefit of using CeO_2 nanoshapes is that the surface facets of each shape are well defined, as the CeO_2 octahedra have (111) surface terminations, the CeO_2 cubes expose (100) surface facets, while the CeO_2 rods have (100) and (110) surface facets. Therefore, this allows us to not only determine the influence of the support oxide surface structure on the active metal, but also facilitate identifying structure-activity relationships of reactions under investigation.

By tuning the ALD deposition conditions, such as low deposition temperature (120 °C) with a short dose time (5s), we can control the Pt loading on the CeO_2 nanoshapes down to a few ppm [2]. At these ultra-low loadings, the Pt/CeO_2 catalysts after calcination in air at 350 °C have only isolated Pt ions dispersed on CeO_2 surface as confirmed by scanning transmission electron microscopy (STEM) and diffuse reflectance infrared Fourier transform spectroscopy (DRIFTS). It is shown that mild reduction temperature 150 °C reduces initial Pt ions and enables isolated Pt atoms to stay intact on the three different ceria facets. In contrast, at a reduction temperature of 350 °C, the Pt-ceria interactions are distinctively dependent on the specific ceria surface facet, and lead to three types of Pt ensembles, including Pt single atoms on ceria (111) facet, flat Pt clusters on oxygen vacancy rich (110) facets and Pt nanoparticles on (100) facet.

The Pt flat cluster structures which formed on the ceria rod support show significant activity enhancement in the CO oxidation reaction even at the ppm level Pt loading. Due to the ultra-low Pt loading, the turnover frequency of the Rod-5.5ppm catalysts in the current study is orders of magnitude higher than those commonly reported in the literature [3-5]. In situ characterization using (DRIFTS) demonstrates that the electron-deficient Pt species on the ceria octahedra result in stronger CO bonding, and this together with very limited reverse oxygen spillover from the CeO_2(111) surface to Pt yield a low activity in the CO oxidation reaction. The highly reducible CeO_2(100) surface of the cubes instead triggers Pt agglomeration into nanoparticles which leads to lower Pt atomic efficiency and poor stability.

This work demonstrates that by utilizing ultra-low loadings of precious metals together with well-defined oxide supports atomic level information on the influence of support structure on heterogeneous catalysts can effectively be probed and this offers guidance for future catalyst design and synthesis.

Reference List

[1] Wu, Z.;Li, M.;Howe, J.;Meyer, H. M., 3rd; Overbury, S. H. Probing defect sites on CeO2 nanocrystals with well-defined surface planes by Raman spectroscopy and O2 adsorption. *Langmuir* **2010**, *26*, 16595-16606.

[2] Song, B.;Choi, D.;Xin, Y.;Bowers, C. R.; Hagelin-Weaver, H. Ultra-Low Loading Pt/CeO2 Catalysts: Ceria Facet Effect Affords Improved Pairwise Selectivity for Parahydrogen Enhanced NMR Spectroscopy. *Angew. Chem. Int. Ed. Engl.* **2021**, *60*, 4038-4042.

[3] Wang, H.;Liu, J. X.;Allard, L. F.;Lee, S.;Liu, J.;Li, H.;Wang, J.;Wang, J.;Oh, S. H.;Li, W.;Flytzani-Stephanopoulos, M.;Shen, M.;Goldsmith, B. R.; Yang, M. Surpassing the single-atom catalytic activity

limit through paired Pt-O-Pt ensemble built from isolated Pt1 atoms. *Nat Commun* **2019**, *10*, 3808.

[4] Jan, A.;Shin, J.;Ahn, J.;Yang, S.;Yoon, K. J.;Son, J. W.;Kim, H.;Lee, J. H.; Ji, H. I. Promotion of Pt/CeO2 catalyst by hydrogen treatment for low-temperature CO oxidation. *Rsc Advances* **2019**, *9*, 27002-27012.

[5] Gatla, S.;Aubert, D.;Agostini, G.;Mathon, O.;Pascarelli, S.;Lunkenbein, T.;Willinger, M. G.; Kaper, H. Room-Temperature CO Oxidation Catalyst: Low-Temperature Metal–Support Interaction between Platinum Nanoparticles and Nanosized Ceria. *ACS Catalysis* **2016**, *6*, 6151-6155.

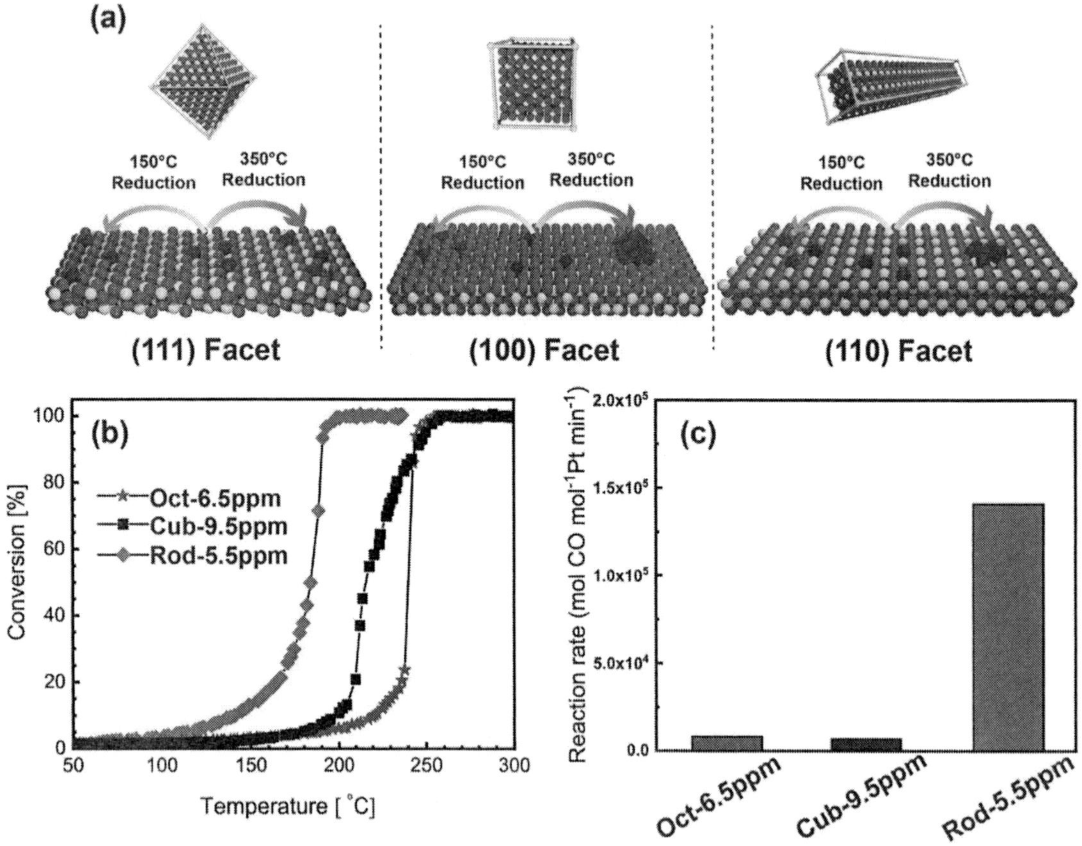

Figure 1. (a) A scheme illustrating different behaviors of Pt single atoms over reduction treatment at temperature 150 °C and 350 °C, (b) The CO oxidation activity over Rod-5.5ppm, Oct-6.5ppm and Cub-9.5ppm. (c) Reaction rate at 190 °C. CO oxidation reaction conditions: 1% CO, 10% O_2, 100 ml/min, 10 mg of catalyst, 10°C/min.

621598 Investigating Colloidal PbSe Quantum Dots As a Room Temperature Operating Material for Gamma Ray Spectroscopy

Abstract:

Semiconductor diode radiation detectors serve as the industry standard for gamma ray spectroscopy. Most of these are fabricated with high-purity germanium (HPGe) single crystals, where energy resolution below 1% can be obtained for 662 keV gamma rays (Cs-137). Despite the maturity of this technology, it has several drawbacks. HPGe detectors need to be operated at liquid nitrogen temperatures and are costly to produce due to the required high purity of the single crystals. Developing a gamma ray semiconductor detector that can operate at room temperature, while matching industry standard resolution has been a challenge. Recently it was proposed that PbSe colloidal quantum dots (QDs) can be used to make low-cost gamma ray detectors, which allow gamma ray spectroscopy to be performed. PbSe has an exciton Bohr radius of 46 nm, which allows the exploitation of quantum confinement effects to tune electronic and optical properties. Synthetic methods to control the diameter and composition of PbSe colloidal QDs have been demonstrated where the electronic and optical properties can be precisely tuned. PbSe QDs also exhibit multiple exciton generation (MEG) through impact ionization (II) where more than one exciton can be produced by one interaction with ionizing radiation. These properties make PbSe QDs a strong candidate as a room temperature gamma ray detector. In this study, we have developed a microwave assisted synthetic protocol for PbSe QDs. The PbSe QDs were characterized by transmission electron microscopy, energy dispersive X-ray spectroscopy, UV-Vis-NIR spectroscopy, Fourier transform infrared spectroscopy, and X-ray diffraction. We found that the QDs were spherical and the diameters could be controlled to between 4-8 nm. QD based devices were fabricated with as synthesized, or ligand exchanged material, and electronically characterized. Their response to gamma radiation was monitored using a linear pulse circuit and multiple channel analyzer.

632443 Effect of Surfactant in Continuous Milllifluidic Green Synthesis of One-Dimensional Silver Nanostructures Using Tannic Acid

Abstract:

Continuous flow processes for the synthesis of nanomaterials have been regarded as more versatile techniques compared to conventional batch processes as they allow better control over the synthesis process, and have a stronger potential for scale-up. Increased mass and heat transfer facilitated by their high surface-to-volume ratio with the ability to allow accurate flow control, and being inherently safer are among their other benefits. Millifluidic flow-reactors, compared to microfluidic chips, are typically more advantageous for particularly the synthesis of metallic nanostructures due to their significantly-lower cost, simpler design and installment procedures, and more resistance to fouling.

However, although millifluidic reactors are more resistant to fouling compared to microfluidic chips, they are still prone to fouling, which hinders the effective production of nanomaterials and affects the size and morphology of the synthesized nanostructures. One of the methods to overcome this issue is to entrap the movement of droplets in order to minimize the interaction of synthesis species with the reactor inner wall. This allows a more accurate control over the synthesis process that is being confined within tiny milli-droplets. This is often facilitated by a secondary-phase flow (e.g. silicone oil) that is inert, and immiscible with water. However, it is highly likely that the synthesized nanoparticles migrate to the water-oil boundary, and resulting in less nanoparticles being present in the aqueous phase, thus limiting the production rate. One particular way to address this issue is to use a non-ionic surfactant that can prevent the diffusion of nanoparticles to the oil-water boundary. In this work, we report the synthesis of silver nanowires (AgNWs) via the millifluidic system by using tannic acid as both the reducing and capping agent, and silicone oil as the secondary-phase flow. We investigate how the synthesis of silver nanowires is affected both with and without the presence of the surfactant (triton X-100). In addition, different flow patterns, flowrates, and other conditions (i.e. pH, tannic acid concentration, and silver precursor concentration) are investigated to observe how the yield of silver nanowires is affected.

The characterization tools used to analyze the resultant silver nanostructures are Ultraviolet–visible Spectroscopy (UV-vis), Scanning Electron Microscopy (SEM), Transmission Electron Microscopy (TEM), and Energy Dispersive X-Ray (EDX). The yield of synthesized AgNWs in continuous millifluidic reactor using surfactant is also compared with batch reactor.

631536 Solving the Population Balance Equation: A Novel Quadrature Method

Abstract:

Moment models are based on specific, discrete or continuous, parametric representations for the probability density function (pdf) and involve/entail the recovery of the underlying pdf when a finite set of its moments is given; this is the essence of the classical (truncated) *moment problem*. Several moment-based approaches for pdf-reconstructions have been proposed in the mathematical literature [1-4], and in particular various adaptations have been formulated as specialized solution methods for the population balance equations (PBEs). Among the moment closure methods for PBEs the quadrature-based moment methods (QBMM), which are based on the *Gauss quadrature* assumption, have gained ground [5-6]. In particular, the quadrature method of moments (QMOM), introduced by McGraw (1997) [7] is based on the basic assumption that the underlying pdf is represented by a weighted sum of delta functions, in which the quadrature nodes and weights are determined by inverting a finite set of its moments.

However, numerical instabilities may occur when nodes lie outside the support interval, or their values are close, or when weights have negative values, leading to solutions that are not physically realizable [8-10]. None of the current methods are capable of dealing with these problems. In this work we present a new method based on Gauss quadrature assumption which refines the standard QMOM formulations and guarantees that the nodes are always lie in the support and the reconstruction of the distribution preserves positivity. We discuss cases where the standard QMOM fails and demonstrate the stability of the new method.

References

[1] Mead, L. R., & Papanicolaou, N. (1984). Maximum entropy in the problem of moments. *Journal of Mathematical Physics*, *25*(8), 2404-2417.

[2] Athanassoulis, G. A., and P. N. Gavriliadis. "The truncated Hausdorff moment problem solved by using kernel density functions." Probabilistic Engineering Mechanics 17, no. 3 (2002): 273-291.

[3] John, V., I. Angelov, A. A. Öncül, and D. Thévenin. "Techniques for the reconstruction of a distribution from a finite number of its moments." *Chemical Engineering Science* 62, no. 11 (2007): 2890-2904.

[4] Gavriliadis, P. N., & Athanassoulis, G. A. (2012). The truncated Stieltjes moment problem solved by using kernel density functions. *Journal of Computational and Applied Mathematics*, *236*(17), 4193-4213.

[5] Yuan, C., Laurent, F., & Fox, R. O. (2012). An extended quadrature method of moments for population balance equations. *Journal of Aerosol Science*, *51*, 1-23.

[6] Madadi-Kandjani, E., & Passalacqua, A. (2015). An extended quadrature-based moment method with log-normal kernel density functions. *Chemical Engineering Science*, *131*, 323-339.

[7] McGraw, R. (1997). Description of aerosol dynamics by the quadrature method of moments. *Aerosol Science and Technology*, *27*(2), 255-265.

[8] Li, D., Li, Z., & Gao, Z. (2019). Quadrature-based moment methods for the population balance equation: An algorithm review. *Chinese Journal of Chemical Engineering*, *27*(3), 483-500.

[9] Vikas, V., Hauck, C. D., Wang, Z. J., & Fox, R. O. (2013). Radiation transport modeling using extended quadrature method of moments. *Journal of Computational Physics*, *246*, 221-241.

[10] Su, J., Gu, Z., Li, Y., Feng, S., & Xu, X. Y. (2008). An adaptive direct quadrature method of moment for population balance equations. *AIChE journal*, *54*(11), 2872-2887.

630321 Towards a Generic Model for Twin-Screw Wet Granulation: Linking Material Properties to PBM Parameters.

Abstract:

Recently, the pharmaceutical industry has changed to produce solid oral dosages from the traditional inefficient and expensive batch production to continuous methods. In this context, the twin-screw wet granulator (TSWG) has become a promising wet granulation technique for continuous manufacturing of solid dosage forms due to its advantages, such as design flexibility, short residence time, and throughput range [1]. In parallel, population balance modeling (PBM) has emerged as a powerful model-based tool not only to predict the output of the desired property but also to increase the understanding and the enhancement of continuous twin-screw granulation processes [2]–[4]. One-dimensional PBM is still widely used in pharmaceutical manufacturing that primarily tracks only one particle property at a time, partly due to the lack of suitable equipment capable of measuring two or more properties simultaneously and the lack of suitable multi-dimensional kernels that have taken into account the specific process or the available experimental data [5], [6].

Several models to simulate the twin-screw wet granulation process have been published [1], [2], [5], [7]–[12]. However, a complete or more generic model involving material properties, process parameters and screw configurations is, to the authors' knowledge, not yet available in literature [13]. This work seeks to take the first steps to address the development of a generic model, which from an industrial perspective is desirable to reduce the number of experiments required to expedite the time to market for new products. A one-dimensional compartmental PBM for the twin-screw wet granulation process is presented to simulate the particle size distribution. Pure aggregation in the wetting zone and a combination of aggregation and breakage in the kneading zones are assumed as the primary phenomena in the granulator. The aggregation kernel in the wetting zone corresponds to a modification of the aggregation kernel originally presented in the work by Van Hauwermeiren et al., 2018 [14]. In particular, a strategy is applied to reduce the amount of unknown parameters of the model in each compartment using patterns detected from the collected experimental data reducing the complexity of the calibration process.

The PBM was calibrated for three different Active Pharmaceutical Ingredients (APIs) of different nature, both, hydrophilic and hydrophobic, at low and high concentrations. Then, six formulations were studied at different granulator process conditions (the granules analyzed in this study were produced in the high shear twin-screw wet granulator module of the ConsiGma-25 from GEA Pharma Systems, ColletteTM, Wommelgem, Belgium, continuous line). Subsequently, Partial Least Squares (PLS) models were built using the calibration results to predict the PBM parameters from the material properties and liquid-to-solid ratio conditions as inputs. This generic PBM for TSWG will greatly reduce the amount of material needed for experimentation and it can reduce the time-to-market and experimental product development costs for a new product.

Acknowledgments:

Pfizer Inc.

Janssen Pharmaceutica NV

UCB

References:

[1] A. Kumar *et al.*, "Linking granulation performance with residence time and granulation liquid distributions in twin-screw granulation: An experimental investigation," *Eur. J. Pharm. Sci.*, 2016.

[2] D. Barrasso and R. Ramachandran, "Qualitative Assessment of a Multi-Scale, Compartmental PBM-DEM Model of a Continuous Twin-Screw Wet Granulation Process," *J. Pharm. Innov.*, vol. 11, no. 3, pp. 231–249, 2016.

[3] A. D. Mcguire *et al.*, "twin-screw granulation Part 1 : Model description New Museums Site," no. 191, 2017.

[4] L. G. Wang, S. U. Pradhan, C. Wassgren, D. Barrasso, D. Slade, and J. D. Litster, "A breakage kernel for use in population balance modelling of twin screw granulation," *Powder Technol.*, vol. 363, pp. 525–540, 2020.

[5] H. Y. Ismail, M. Singh, A. B. Albadarin, and G. M. Walker, "Complete two dimensional population balance modelling of wet granulation in twin screw," *Int. J. Pharm.*, vol. 591, no. April, p. 120018, 2020.

[6] M. Singh, A. Kumar, S. Shirazian, V. Ranade, and G. Walker, "Characterization of Simultaneous Evolution of Size and Composition Distributions Using Generalized Aggregation Population Balance Equation," *Pharmaceutics*, vol. 12, no. 12, p. 1152, Nov. 2020.

[7] D. Barrasso, S. Walia, and R. Ramachandran, "Multi-component population balance modeling of continuous granulation processes: A parametric study and comparison with experimental trends," *Powder Technol.*, vol. 241, pp. 85–97, 2013.

[8] A. D. Mcguire *et al.*, "twin-screw granulation Part 1 : Model description New Museums Site," *Chem. Eng. Sci.*, vol. 188, no. 191, pp. 18–33, 2017.

[9] D. Van Hauwermeiren *et al.*, "On the modelling of granule size distributions in twin-screw wet granulation: calibration of a novel compartmental population balance model," *Powder Technol.*, pp. 116–125, 2018.

[10] S. Shirazian, H. Y. Ismail, M. Singh, R. Shaikh, D. M. Croker, and G. M. Walker, "Multi-dimensional population balance modelling of pharmaceutical formulations for continuous twin-screw wet granulation: Determination of liquid distribution," *Int. J. Pharm.*, vol. 566, no. June, pp. 352–360, 2019.

[11] M. Singh, G. Kaur, T. De Beer, and I. Nopens, "Solution of bivariate aggregation population balance equation: a comparative study," *React. Kinet. Mech. Catal.*, vol. 123, no. 2, pp. 385–401, 2018.

[12] L. G. Wang, S. U. Pradhan, C. Wassgren, D. Barrasso, D. Slade, and J. D. Litster, "A breakage kernel for use in population balance modelling of twin screw granulation," *Powder Technol.*, vol. 363, 2020.

[13] P. Thapa, J. Tripathi, and S. H. Jeong, "Recent trends and future perspective of pharmaceutical wet granulation for better process understanding and product development," *Powder Technol.*, vol. 344, pp. 864–882, 2019.

[14] D. Van Hauwermeiren *et al.*, "On the modelling of granule size distributions in twin-screw wet granulation : Calibration of a novel compartmental population balance model," *Powder Technol.*, vol. 341, pp. 116–125, 2018.

630419 High Speed Friction Grinding of Thermoplastics Assisted By Food Gums

Abstract:

High speed friction grinding (fibrillation technology) has been used to grind plant in water but never been explored for grinding of thermoplastics. Its most common use is for converting cellulose pulp into nanofibrillated cellulose. Thermoplastics like polylactic acid (PLA), low and high density polyethylene (LDPE and HDPE) and polypropylene (PP) are hydrophobic and elastic in nature that make them difficult to fracture into fine powders at room temperature. This study examines a new concept involving use of food gums for grinding of thermoplastics into finely ground particles in an environmental friendly and cost-effective way. Low concentrations of food gums (like 0.5% guar gum in water) made such grinding possible as they increased the viscosity of water which reduced the settling of plastics and increased the residence time through the grinder. Tensile, flexural and impact strengths of the plastics were studied and higher grinding efficiency of PLA could be explained by its low elongation-at-break compared to LDPE, HDPE and PP. The microplastics (2000-45 μm) were studied for mass and particle size distributions and by SEM, 13C CP/MAS NMR, FTIR, DSC and TGA. Viscosity of guar gum and contact angles were measured. This new technology can produce finely ground microplastics (710–45 μm) for a variety of applications such as additive manufacturing, plastic coating on metal substrates, biomedical devices, upcycling and recycling.

Reference:

Bhagia, S, Gallego, NC, Hiremath, N, et al. Fine grinding of thermoplastics by high speed friction grinding assisted by guar gum. *J Appl Polym Sci*. 2021;e50797. https://doi.org/10.1002/app.50797

627046 Unsteady-State Simulation of Cement Milling in Closed-Circuit Operation: A Fully Coupled Ball Mill-Air Classifier Model

Abstract:

As an important unit operation, ball milling is widely used in numerous industries, including the cement industry. Using a ball mill, which is a rotating horizontal tube containing grinding media as balls, is inefficient because the process generates losses in heat, vibration, and sound energy [1]. Nevertheless, the majority of cement clinker mills are operated using ball milling technology [2]. As the shell of a ball mill is rotated, its grinding balls move in a cataracting mode, hitting on the powder bed, and cement clinker particles are broken down to smaller sizes. In closed-circuit mode, ball milling operation and external air classification are coupled (see Fig. 1). A dynamic air classifier equipped with a rotor (the rotating part) driven by a separate motor and stationary guide vanes classifies discharged powder from a mill into a fine product stream (final product) and a coarse reject stream. The classification area of an air classifier is the space between the rotor cage and the guide vanes. Particles entering this space are subjected to centrifugal, drag, and gravitational forces. Depending on the magnitude and direction of these forces, the particles move along different trajectories. The desired product size can be adjusted by changing airflow rate and/or rotor speed [3].

Several studies have examined the impacts of process parameters, including the selection of ball sizes, on milling performance (e.g., [4-6]). However, experimental studies intended to identify optimum operating parameters were performed under small-scale batch ball milling; industries operate ball mills in continuous mode. Past experimental research has also indicated that a small ball more effectively breaks down fine particles, whereas a large ball more effectively breaks coarse particles. Thus, in a poly-dispersed (natural) feed where coarse and fine particles coexist, a mixture of ball sizes can be expected to outperform a single ball size. In view of breakage kinetics, operating with an excessive number of ball sizes does not necessarily yield finer products. As experimentally observed [7], a binary mixture of ball sizes can generate either finer or coarser products than those of a quaternary size mixture, depending on the selected top ball size. On the basis of the actual operation of industrial mills [8,9], the selection of up to five ball sizes is common in large mills. However, a caveat in opting for a binary mixture of considerably differing ball sizes is that the wear rate of balls may be high and thereby increase operating costs.

As mentioned earlier, the desired product size can be controlled by adjusting airflow rate and rotor speed of the external air classifier [3]. A decrease in airflow rate (or an increase in rotor speed) yields finer products, but it elevates the mass flow rate of a coarse reject stream. A critical issue is that these operation strategies are valid only when an air classifier is large enough to handle the powder discharged by a ball mill. The classifier cannot be operated at too low an airflow rate (or too fast a rotor speed) because it is overloaded by an excessive mass flow rate of a coarse reject stream. Such overload, in turn, sharply diminishes classification performance, thus shutting down operation. In other words, a closed-circuit ball mill operates at optimum under a certain coarse reject rate [10]. Recently, the classification function, also known as the Tromp curve (the variable Tromp curve, hereafter), was empirically developed to factor in the dust load carried by an air classifier [11]. Although the variable Tromp curve was proposed on the grounds of industrial cement milling data, its corresponding model has never been incorporated into a closed-circuit mill—a challenging task because of an increase in the nonlinearity of the coupled system. Given the aforementioned complexity during milling-classification, uncovering optimum operating conditions is difficult and costly. Alternatively, a simulation tool can illuminate the interplay between milling and classification as well as clear the way for pinpointing optimum operation at minimal cost.

The population balance model (PBM) has been used to simulate particulate processes, thereby enabling the quantitative analysis of the spatio-temporal evolution of particle size distribution (PSD) during milling. Simulators discussed in the open literature and even the commercial software for closed-circuit dry milling simulators, such as JKSimMet [12], are steady-state simulators. Unlike other simulators, the true unsteady-state simulator (TUSSIM) solves for temporal evolution of the PSD within the mill and simulates the coupled ball mill-air classifier system during full-scale dry ball milling in the closed-circuit mode (see Fig. 2). TUSSIM is based on the solutions of a cell-based PBM [13], which was augmented with the variable Tromp curve [11]. It consists of a set of differential-algebraic equations (DAEs) and accounts for realistic finite mixing, internal particle separation caused by diaphragms, external particle separation caused by air classifiers, a combination of ball sizes, ball size distribution, and the application of classifying liners. Using the TUSSIM, we explored the impact of ball mixtures, as opposed single-sized balls, the action of classifying liner, and rotor speed–air flow rate of the external air separator on the product PSD, recirculation ratio, dust load to the air classifier, etc.

Our simulation results suggest that ball mixtures generated finer products than a single ball size, in correspondence with actual industry operations [1]. The use of a classifying liner yielded finer products as opposed to those produced through uniformly mixed ball sizes. In consideration of particle classification, reduced airflow rate (or augmented rotor speed) in the air classifier yielded finer product PSD and, hence, a higher coarse reject rate. These outcomes are congruent with the findings of previously conducted experiments [3]. Even more interestingly, under a further decrease in airflow rate (or a further increase in rotor speed), TUSSIM can detect the failure of the system when the air classifier is overloaded (high dust loading). Overall, we have demonstrated that the successful development of TUSSIM, a PBM-based simulation tool, will enable the industry to improve the closed-circuit ball milling operations through enhanced process understanding and especially the complex coupling of ball milling with air classification system.

References

[1] W.H. Duda, Cement Data Book, Volume One: International Process Engineering in the Cement Industry, 3rd ed. Bauverlag GmbH, Wiesbaden und Berlin, 1985.

[2] D. Xu, Y. Cui, H. Li, K. Yang, W. Xu, Y. Chen, On the future of Chinese cement industry, Cem. Concr. Res. 78 (2015) 2-13.

[3] L. Gao, Y. Yu, J. Liu, Study on the cut size of a turbo air classifier, Powder Technol. 237 (2013) 520-528.

[4] L.G. Austin, K. Shoji, P.T. Luckie, The effect of ball size on mill performance, Powder Technol. 14 (1976) 71-79.

[5] V. Deniz, A study on the specific rate of breakage of cement materials in a laboratory ball mill, Cem. Concr. Res. 33 (2003) 439-445.

[6] N. Kotake, K. Suzuki, S. Asahi, Y. Kanda, Experimental study on the grinding rate constant of solid materials in a ball mill, Powder Technol. 122 (2002) 101-108.

[7] S. Cayirli, Influences of operating parameters on dry ball mill performance, Physicochem. Probl. Miner. Process. 54 (2018) 751-762.

[8] O. Genc, Optimization of an industrial scale open circuit three-compartment cement grinding ball mill with the aid of simulation, Int. J. Miner. Process. 154 (2016) 1-9.

[9] V. Ghalandari, A. Iranmanesh, Energy and exergy analyses for a cement ball mill of a new generation cement plant and optimizing grinding process: a case study, Adv. Powder Technol. 31 (2020) 1796-1810.

[10] J. Yardi, Separator technology, World Cement 36 (2005) 87-104.

[11] O. Altun, H. Benzer, Selection and mathematical modelling of high efficiency air classifiers, Powder Technol. 264 (2014) 1-8.

[12] S. Schwarz, J.M. Richardson, Modeling and simulation of mineral processing circuits using JKSimMet and JKSimFloat, SME Annu. Meet., 2013.

[13] N. Muanpaopong, R. Davé, E. Bilgili, Powder Technol., in submission.

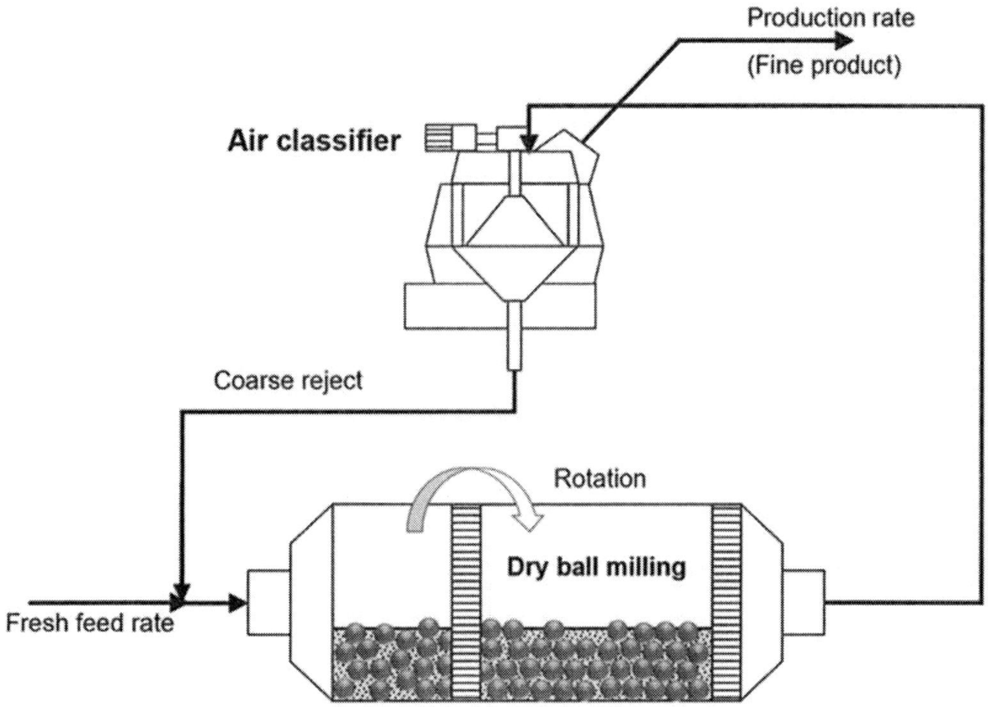

Fig. 1. Sketch of the closed-circuit, continuous dry milling process incorporating intermediate and discharge diaphragms and air classifier.

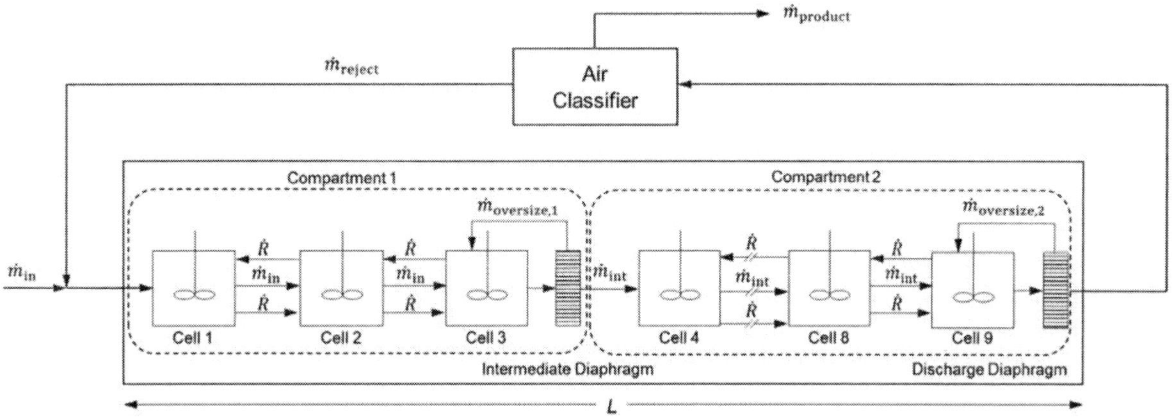

Fig 2. Schematic model of the closed-circuit ball mill represented as a series of nine perfectly mixed cells with intermediate and discharge diaphragms and an external air classifier. The cells exchange particles at the rate $\dot{R}$. The mill has effective length L.

626705 A Hybrid Model to Predict the Formulation Dependent Granule Growth in a Bi-Component Wet Granulation Process

Abstract:

The wet granulation process is often a complex unit operation used in various particle processing industries. The complexity of this process lies in understanding the effects of equipment design, operating conditions, and material properties on the granule product quality. Significant knowledge has been accumulated on the population balance modeling of the wet granulation process where the impact of operating conditions on granule product properties was studied [3,2,1,6]; however, there is comparatively less insight on the impact of material properties. The constituent materials involved in the process can vary in properties such as shape, size, hydrophobicity, and strength. This material variability impacts the granulation mechanisms and affects the critical granule properties such as particle size distribution, porosity, microstructure, and content uniformity. Thus, quantifying the effects of material variability on granules is critical for a holistic understanding of the wet granulation process. In this study, a bicomponent wet granulation process is considered, and a modeling framework to predict the impact of material hydrophobicity is proposed.

Although many drug products involve materials with wettability differentials, there is a lack of understanding of how the hydrophobic behavior of materials affects the wet granulation process and granule product quality. To study the impact of wettability/hydrophobicity on granulation, two formulations were considered (I: Ibuprofen, and Microcrystalline cellulose (MCC-101), II: Acetaminophen, and MCC-101) with a wide difference in contact angle. Wet granulation results showed that the compositional distribution of components among granules of different sizes (i.e., content uniformity) is impacted by the granule growth mechanism. Ibuprofen formulation favored viscous force dominant granule growth, and acetaminophen formulation favored capillary force dominant granule growth. Due to viscous force dominant growth, ibuprofen formulation produced granules with a uniform compositional distribution of components among granules of different sizes. Capillary force dominant granule growth of acetaminophen formulation leads to weaker granules resulting in a non-uniform distribution of components among granules of different sizes [5]. Current population balance models lack the ability to capture this phenomenon.

Based on the experimental observations, a hybrid modeling framework was developed for predicting the granule properties in a bi-component wet granulation system with components of differing hydrophobicities(Figure 1). First, a random forest method is used for predicting the probability of nucleation mechanism (immersion and solid-spread) depending upon the formulation hydrophobicity. The predicted nucleation probability is used to determine the aggregation rate as well as the initial particle distribution in the population balance model. The aggregation process was modeled as Type-I: Sticking aggregation and Type-II: Deformation driven aggregation [4]. In Type-I, the capillary force dominant aggregation mechanism is represented by the particles sticking together without deformation. In the case of Type-II, the particle deformation causes an increase in the contact area, representing a viscous force dominant aggregation mechanism. The choice between Type-I and II aggregation is determined based on the difference in nucleation mechanism that is predicted using the random forest method. The model was optimized and validated using the granule content uniformity data obtained from the experimental studies. The proposed framework predicted content non-uniform behavior for formulations that favored immersion nucleation and uniform behavior for formulations that favored solid-spreading nucleation.

References

[1] Chaudhury, A., Armenante, M.E., Ramachandran, R., 2015. Compartment based population balance modeling of a high shear wet granulation process using data analytics. Chemical Engineering Research and Design 95, 211–228.

[2] Chaudhury, A., Wu, H., Khan, M., Ramachandran, R., 2014. A mechanistic population balance model for granulation processes: effect of process and formulation parameters. Chemical Engineering Science 107, 76–92.

[3] Darelius, A., Brage, H., Rasmuson, A., Bj¨orn, I.N., Folestad, S., 2006. A volume-based multidimensional population balance approach for modelling high shear granulation. Chemical Engineering Science 61, 2482–2493.

[4] Goodson, M., Kraft, M., Forrest, S., Bridgwater, J., 2004. A multi-dimensional population balance model for agglomeration, in: PARTEC 2004—International Congress for Particle Technology.

[5] Muthancheri, I., Ramachandran, R., 2020. Mechanistic understanding of granule growth behavior in bi-component wet granulation processes with wettability differentials. Powder Technology 367, 841–859.

[6] Pandey, P., Tao, J., Chaudhury, A., Ramachandran, R., Gao, J.Z., Bindra, D.S., 2013. A combined experimental and modeling approach to study the effects of high-shear wet granulation process parameters on granule characteristics. Pharmaceutical development and technology 18, 210–224.

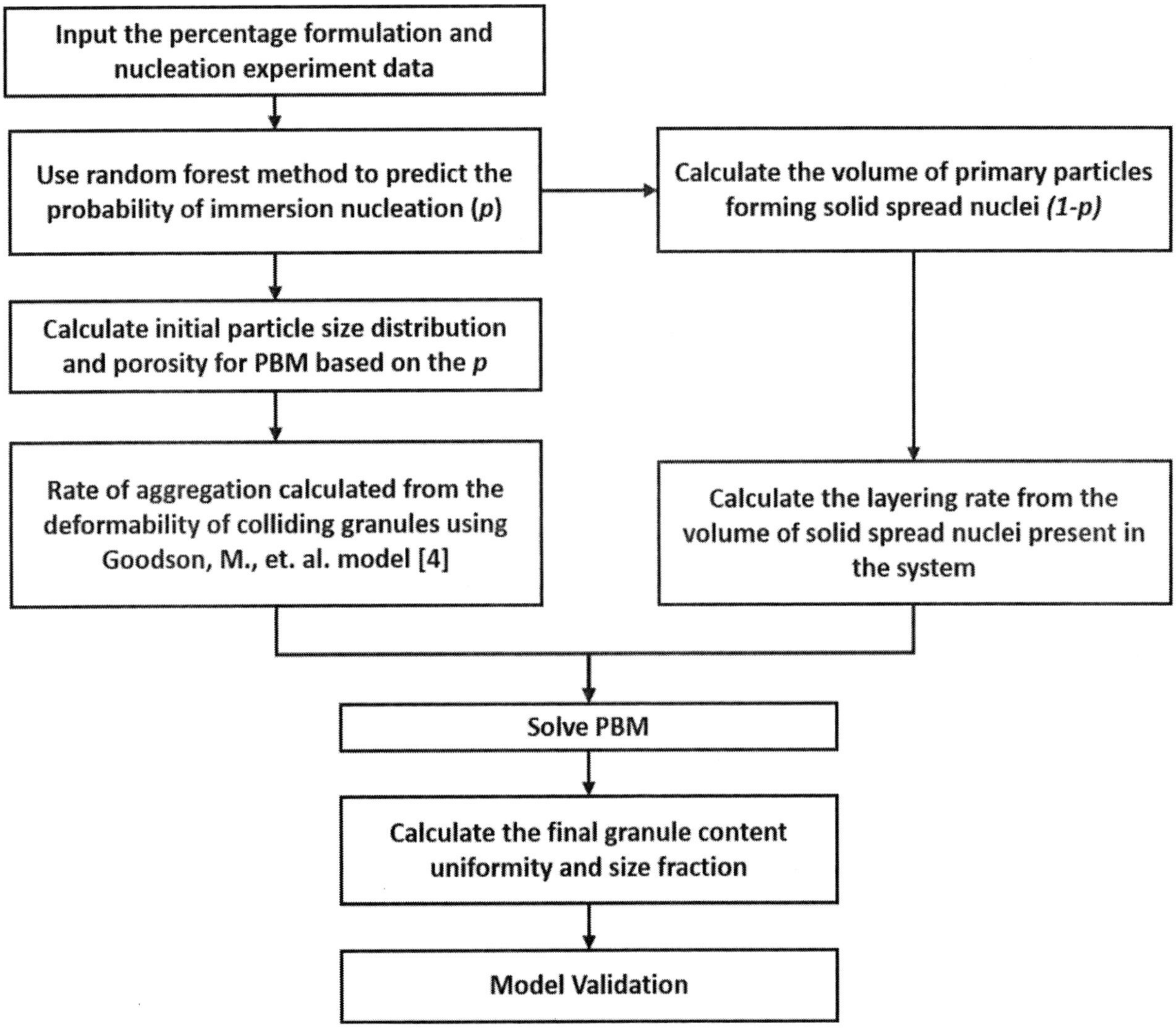

624164 Sonofragmentation of Two-Dimensional Plate-like Crystals: Population Balance Modelling and Experimental Validation

Abstract:

The use of multidimensional population balance equations in predicting ultrasound-assisted sonocrystallization processes is curbed by the unavailability of physically motivated macroscopic mathematical kernels. This study develops a comprehensive population balance modelling of the ultrasound assisted sonofragmentation of thin rectangular plate-type pyrazinamide crystals. A new bivariate breakage selection function is presented, which takes care of the time dependence in particle selections along with the size dependence. Furthermore, a new mathematical formulation of the bivariate breakage distribution function is introduced to predict the outcomes of the sonofragmentation experiments of Bhoi et al. (Chem. Eng. Sci. 2019, 203, 12-27). Finally, the results obtained from solving the developed population balance model are successfully validated against the experimental results. The developed population balance model accurately captures the effects of variations in ultrasonic amplitude and sonication time.

619869 A Molecular Dynamics Study of Interparticle Interactions in Fresh Cement Pastes

Abstract:

A Molecular Dynamics Study of Interparticle Interactions in Fresh Cement Pastes

Juan Pablo Gallo-Molina[1,2], Adilson Alves de Freitas[3], Karel Lesage[2], José Nuno Canongia Lopes[3] and Ingmar Nopens[1]

[1]BIOMATH, Department of Data Analysis and Mathematical Modelling, Ghent University, Ghent, Belgium.

[2]Magnel-Vandepitte Laboratory, Department of Structural Engineering and Building Materials, Ghent University, Ghent, Belgium.

[3]Centro de Química Estrutural, Instituto Superior Técnico, Universidade de Lisboa, Lisbon, Portugal.

Due to the fact that it is one of the most produced materials in the world, its critical importance for infrastructure development and the environmental issues associated with the large masses involved, concrete has been attracting increasing research attention since the last decades[1]. For hardened concrete, civil engineers already possess a toolset that allow them to satisfactorily predict properties of interest (e.g. deformation, stress, etc.) as a function of material characteristics. However, when this material is encountered in a fluid state, the picture is completely different: only some empirical tests can relate some material variables (e.g. concrete composition) with relevant properties (i.e. flow speed, flow spread, etc.). Although this practice is longstanding, the increasing complexity of concrete mixtures - introduced by the growing use of mineral additions, chemical admixtures and non-traditional cement chemistries - has invoked the need for developing mechanistic models with predictive capabilities.

Typical concrete suspensions exhibit very broad polydispersity, with size ranges encompassing several orders of magnitude. Furthermore, composition and crystalline structures vary with both size and source. It follows that rheology is not controlled by a single mechanism, but rather by several competing ones. Indeed, hydrodynamic, van der Waals and electrostatic forces all play a role in determining flow behavior. Likewise, steric interactions become relevant if chemical admixtures are used and the aforementioned hydration reactions grow in importance after the first hours following water addition [2]. The population balance framework has been identified as a valuable tool that is capable of reflecting the emerging nature of rheology from the transient evolution of a given cement suspension microstructure[3]. However, to increase its effectivity, precise knowledge of the nature of colloidal interactions is needed, as they play a vital role in determining the values of aggregation efficiencies used by population balances models.

Considering this, we present in this work a molecular dynamics model focused on the main phases found in cement clinkers[4]. The objective of the modelling effort is to better understand the behavior of interparticle interactions in these systems since, as suggested before, they have a direct impact on agglomeration dynamics and subsequently rheology. We investigate the presence of interaction phenomena that cannot usually be assessed with classical formulations. Moreover, the effect of crystalline structure as well as that of interparticle distance is analyzed. The final aim is to gain relevant information for the construction of a more accurate representation of agglomeration/breakage dynamics in cement pastes.

References

[1] R. J. Flatt, N. Roussel, and C. R. Cheeseman, "Concrete: An eco material that needs to be improved," *J. Eur. Ceram. Soc.*, vol. 32, no. 11, pp. 2787–2798, 2012, doi: 10.1016/j.jeurceramsoc.2011.11.012.

[2] N. Roussel, A. Lemaître, R. J. Flatt, and P. Coussot, "Steady state flow of cement suspensions: A micromechanical state of the art," *Cem. Concr. Res.*, vol. 40, no. 1, pp. 77–84, 2010, doi: 10.1016/j.cemconres.2009.08.026.

[3] J. P. Gallo-Molina, K. Lesage, and I. Nopens, "Numerical Validation of a Population Balance Model Describing Cement Paste Rheology," *Materials (Basel).*, vol. 13, no. 5, p. 1249, Mar. 2020, doi: 10.3390/ma13051249.

[4] A. A. Freitas, R. L. Santos, R. Colaço, R. Bayão Horta, and J. N. Canongia Lopes, "From lime to silica and alumina: Systematic modeling of cement clinkers using a general force-field," *Phys. Chem. Chem. Phys.*, vol. 17, no. 28, pp. 18477–18494, 2015, doi: 10.1039/c5cp02823j.

636358 The Lasting Legacy of Prof. John Grace

Abstract:

This session is for invited speakers in honor or Prof. John Ross Grace. Prof. John Grace graduated with a B.E.Sc. in 1965 from the University of Western Ontario, and a Ph.D. from Cambridge University, England, in 1968. For the first 11 years of his distinguished academic career, John was a faculty member in the Chemical Engineering Department at McGill University. In 1979 he then joined the UBC CHBE department as its Head. Among the many prestigious honors he received, Professor Grace was invested as an Officer of the Order of Canada in 2014 for his work in new ways to produce energy with a smaller environmental footprint.

636360 Relationship between the net electrostatic charge inside a fluidized bed and particles accumulation on the column wall

Abstract:

Solids contact charging, known as triboelectrification, could occur in any gas-solid system during solids handling, transport, and processing. However, the extent of charging varies from one system to another depending the operating conditions as well as materials involved. Gas-solid fluidized beds in particular create an ideal environment for triboelectrification to occur because of the continuous collisions among the fluidizing particles and the column wall. Depending on the level of charge generation, operational challenges could arise including particle agglomeration causing deviation from the desired particle properties, accumulation of charged particles on the column wall requiring shutdowns for clean-ups, and electrostatic discharge resulting in a hazardous environment. Hence, much research has been carried out to better understand this electrostatics occurrence and its modeling in fluidized beds [1, 2]. Our latest experimental results show that there is a direct proportionality between the net electrostatic charge of the bulk of the bed particles and the number of particles sticking to the fluidization column wall [3]. Thus, diminishing this net charge will reduce the formation and progression of the column wall fouling. This finding will help in determining means to prevent or reduce fluidizing particle adhesion to the column wall with either manipulation of the fluidization operating conditions or addition of any materials that could counterbalance the existing net charge within the bed.

[1] Mehrani P., Murtomaa M., Lacks D.J., An Overview of advances in understanding electrostatic charge buildup in gas-solid fluidized beds, Journal of Electrostatics, 87, 64-78, 2017.

[2] Chowdhury, F., Ray, M., Sowinski, A., Mehrani, P., Passalacqua, A., A review on the modeling of electrostatic charging in gas-solid flows, Powder Technology, 389, 104-118, 2021.

[3] Taghavivand, M., Mehrani, P., Sowinski, A. Triboelectric effects of continuity additives and a silica catalyst support on polyethylene fluidized bed wall fouling, Chemical Engineering Science, 245, 2021.

636359 Numerical study of particle tracking measurements in fluidized bed reactors

Abstract:

Particle tracking is an important non-invasive technique to measure the solids flow in fluidized bed reactors. In this study, we carried out CFD-DEM simulations to compare different particle tracking methods used in the travelling fluidized bed experimental test campaign led by the UBC team. Different types of tracer particles used in the radioactive particle tracking (RPT) and positron emission particle tracking (PEPT) were simulated to evaluate their impacts on the measured flow fields, e.g., solids velocity and mean solids concentration, and to help understand the significant discrepancy between these two types of measurements observed experimentally. We then evaluated different methods to calibrate the particle tracking measurements for comparison with numerical simulations. It is critical to consistently post-process numerical data and experimental measurements for validation purpose. Some limitations of the trajectory-following experimental techniques were examined. The learning has been applied to the modeling study of a horizontal stirred bed reactor for polypropylene production for which the simulation was validated against RPT measurements in a cold flow unit.

636364 Kinetics & CFD Simulations for Industrial FCC Units

Abstract:

Fluid catalytic cracking (FCC) is an important conversion process in modern refineries, as a major contributor to the gasoline market, and recently an increasingly favored process for producing propylene to feed downstream petrochemical complexes. Detailed analysis of the hydrodynamics in FCC systems has been playing a key role in the development and evolution of the reactor configuration, feeding systems, and riser separation systems, as well as troubleshooting of key process deviations and operational instabilities. Enhancements in computational capabilities have made hydrodynamics assessments using Computational Fluid Dynamics (CFD) tools feasible and affordable for commercial large-scale FCC units with quite reasonable accuracies. This presentation will discuss the importance of generating reliable kinetics using pilot plant data, opportunities that enhancements in CFD computational capabilities will facilitate the development of innovative designs and improvements, while highlighting the role of CFD modeling in the technology development projects at Saudi Aramco.

636361 CFD-DEM simulation of the onset of fluidization for large particles in a bed of fine sand

Abstract:

Due to the unique advantages of fluidized beds for solid mixing and heat transfer, they are widely used for the thermal processing of solid fuel particles. In practice, such systems are composed of non-conventional fuel particles such as coal or biomass that are completely different in shape, size, and density from the bed material such as sand. In such cases, correct understanding of the main fluidization parameters such as the minimum fluidization velocity (U_{mf}) of the bed components plays a key role in the successful design and control of the systems. However, the onset of fluidization of irregular particles is not clear as it may vary depending on the physical properties of the particles, the initial mixing conditions of the conventional and irregular particles, and/or the fraction of the non-conventional particles in the bed.

In this study, we aim at predicting the U_{mf} of large and light biomass particles (objects) mixed with fine sand particles (fines), with the help of coupling the computational fluid dynamics (CFD) and the discrete element method (DEM). To this end, fine particles were primarily simulated using the coarse-grained CFD-DEM, in which a group of fine particles was lumped into a larger numerical parcel. In the conventional CFD-DEM, the CFD cell size is usually 3-4 times larger than the bed particles. This ratio cannot be held when there is a substantial size difference between the bed components as it significantly increases the computational error. To address this issue, we adopted the diffusion-based averaging method and mapped the solid concentration and solid velocity of the object into the CFD cells. The capability of the adopted numerical approach for predicting the motion of the object in a bed of fine particles was verified by comparing the upward and downward velocities of the object with the corresponding experimental data. Further simulations are in progress to predict the U_{mf} of the objects varying in size and density in a bed of sand particles.

636362 Characterization of the main components of fluidization technology applied to thermochemical conversion

Abstract:

Design and scale-up of fluidized beds are significantly impacted by their hydrodynamic features. Cold flow experimentation complemented by CFD simulations are powerful tools for understanding different phases behavior and optimize the design for real operation conditions. This presentation provides instances of hydrodynamic studies utilized for design of processes for thermochemical conversion of Biomass and Petcoke through Chemical Looping Combustion and Pyrolysis. The features of the key elements such as fluidized bed reactors, L-valves and solid fuel injection line are presented through an experimental and numerical approach.

636363 Experimental investigation of wet pharmaceutical granulation using in-situ synchrotron X-ray imaging

Abstract:

Observing and measuring wet granulation processes of pharmaceutical powders is challenging with conventional experimental methods due to their opacity and complex interaction that happens in a short time. In this work, synchrotron in-situ imaging technique was employed to capture the dynamic granulation process with a single drop impacting method. A new parameter, overall movement ratio is introduced to quantitatively describe the movement of the liquid binders in the powder beds. The dynamic interactions between the liquid binder and solid powders were captured via high temporal and spatial resolution X-ray images. Results show that particles with smaller sizes have the tunneling mechanism, and the spreading mechanism occurs with larger particles, as expected. Liquid binder properties including viscosity and contact angle, and operation parameters including droplet release height and volume show significant impacts on the dynamic granulation process. This study demonstrates the potential of using in-situ synchrotron X-ray imaging to quantitatively study the dynamic wet granulation process.

636934 Solids flux profiles in high velocity CFB risers of FCC catalyst particles

Abstract:

Circulating fluidized bed (CFB) risers are widely used in the fluid catalytic cracking (FCC) process. FCC risers typically operate with solids circulation fluxes of 400 – 1400 kg/s.m^2 and superficial gas velocities as high as 15 – 20 m/s. Although extensive work has been conducted on CFB risers over the years, almost all the reported data are for low gas velocities (< 10 m/s) and modest solids circulation rates (< 200 kg/m^{2}s). Limited capacities have often prevented academic studies from achieving higher gas velocity and solids flux conditions. CFB combustor risers operate at much lower gas velocities and solids circulation fluxes, and they utilize Group B solids. The hydrodynamics literature shows that CFB risers consist of a dilute region towards the top and a relatively dense region near the bottom, with the height of the dense region increasing or decreasing depending on the gas superficial velocity and solids circulation rates. The top region has a dilute core in which solids flow rapidly upward, surrounded by a descending dense annulus. The bottom zone has not been widely studied, but the few available studies also suggest an existence of a core-annular flow structure with little or no net downflow of solids in the annulus. This presentation discusses tests conducted in four CFB risers 15 to 25 m in height, all having a diameter of 0.3 m using FCC catalyst particles. Operating superficial gas velocities were between 9 and 17 m/s, and the riser solids fluxes ranged from about 70 to 700 kg/s.m^2. Axial apparent density profiles were determined from pressure gradients measured by differential pressure transmitters. Local solids flux profiles were obtained by solids extraction tubes. High speed video images were acquired in one of the risers. The results revealed local solids flux profiles of various shapes, which depended to a great extent on the superficial gas velocity. They included parabolic profiles with solids downflow near the wall, all upward flow parabolic profiles with maxima in the core, nearly flat profiles, as well as bell-shaped profiles with the highest fluxes near the riser walls. For typical FCC riser operating conditions, the net solids flow direction at all radial locations was found to be upward. This was also confirmed from the high-speed video images.

628734 Comparative Assessment of Spray-Dried Hybrid Nanocrystal–Amorphous Solid Dispersions (HyNASDs) and ASDs

Abstract:

With an increasing number of poorly water-soluble drugs, aqueous solubility has become a critical issue in the pharmaceutical industry. Slow and incomplete release of the poorly water-soluble drugs in the gastrointestinal tract (GI) results in slow absorption and ultimately limited bioavailability. To resolve this issue, among various formulation approaches, drug nanocrystal-based solid dosage forms (nanocomposites) and amorphous solid dispersions (ASDs) have achieved substantial popularity (1–3). In a nanocomposite, drug nanocrystals are usually dispersed in the polymeric matrix as a secondary phase (3), whereas in an ASD, drug is molecularly dispersed in a miscible, amorphous polymer to form a single phase (2). While offering several advantages such as dissolution rate enhancement, improved bioavailability, elimination of food effects, etc., drug nanoparticles have been used in several marketed products (4). However, limited improvement in the bioavailability was observed for drugs with very low aqueous solubility and high drug dose (5). On the other hand, owing to the high apparent (kinetic) solubility, ASDs can generate high extent of supersaturation along with enhanced dissolution rate, which results in bioavailability enhancement of poorly soluble drugs with high therapeutic doses (6). Consequently, drug nanocomposites appear to have limited supersaturation capability, which is the greatest impediment to the bioavailability enhancement and their competitiveness to ASDs.

Being motivated by the use of high polymer concentration, i.e., low drug:polymer mass ratios from 1:1 to 1:9 in drug ASDs (7) unlike the high drug:polymer ratios in drug nanocomposites (3), Rahman et al. (8) have produced nanocomposites with a high polymer concentration (>100% with respect to frug). They used 1:1 to 1:5 drug:polymer mass ratios with/without SDS in the production of griseofulvin (GF) nanocomposites, which is a fast crystallizing poorly soluble drug (9). This approach led to formation of a special class of nanocomposites with notable amorphous drug content (>10%), i.e., hybrid nanocrystal–amorphous solution dispersions (HyNASDs), which showed significant relative supersaturation (~300%) during dissolution. Judicious choice of polymers that have relatively low aqueous viscosities even at high concentrations allowed for preparation of drug nanosuspensions and their spray-drying without any processing issues. To generalize the applicability of HyNASDs, further investigation is required with other drugs and polymer systems.

Being motivated by the earlier work in Rahman et al. (8,10), the main objective of this study is to boost the supersaturation capability of drug nanocomposites in dissolution tests significantly by using low drug:polymer mass ratio (high polymer loading), thus, producing HyNASDs. Unlike Rahman et al. (8,10), HyNASDs were produced using a slow crystallizing poorly water-soluble drug, Itraconazole (ITZ). Additionally, the dissolution performance of HyNASDs and ASDs w/o an anionic surfactant (SDS) was compared in a *head-to-head* fashion. To this end, a wet media milled ITZ suspension and an ITZ solution, containing the same polymer (HPC: hydroxypropylcellulose, Sol: Soluplus, VA64: Kollidon VA6) in deionized water and in dichloromethane (DCM) were spray-dried to prepare ITZ HyNASDs and ITZ ASDs, respectively (see Fig. 1, left and middle panels). To elucidate the impact of SDS during *in vitro* drug release from the HyNASDs, ITZ suspensions with 0.125% SDS and w/o SDS were used. The solid-state characterization of the spray-dried powders as well as physical mixtures (PMs) prepared by blending formulation components without spray drying was performed by X-ray powder diffraction (XRPD) and differential scanning calorimetry (DSC). Release of ITZ from the spray-dried powders was studied using a USP II apparatus coupled with UV-spectroscopy.

Similar XRPD diffractograms (not shown for brevity) to those of the physical mixtures (PMs) were observed for the spray-dried powders produced from aqueous suspension(W)-based feeds confirming that spray-drying of the aqueous milled suspensions led to formation of nanocomposites that are largely crystalline in nature. Interestingly, the diffractograms of the spray-dried powders show significantly reduced peak intensities as compared with their respective PMs, beyond the dilution effect of the polymers.

Wet media milling followed by spray drying led to reduction of crystallinity and formation of notable amorphous (~5–30%) ITZ. In the presence of high polymer loading in the suspensions, amorphization of ITZ seemed to have taken place during the spray drying. Reduction in peak intensity was more pronounced in the case of Sol confirming more amorphous ITZ formation in the Sol formulations than HPC and VA64 formulations. On the other hand, XRPD diffractograms of the spray-dried powders produced from solution-based (S) feeds showed a halo pattern, confirming that ITZ dispersed molecularly into the polymer matrices forming amorphous solid dispersions (ASDs). DSC results confirm the XRPD findings (not shown for brevity).

A Distek 2100C USP II (paddle apparatus) dissolution tester (North Brunswick, NJ, USA) was used to determine ITZ release from the as-received ITZ, spray-dried powders, and the physical mixtures (PMs) with 100 mg equivalent ITZ in 1000 mL 0.1 N HCl solution at 37 °C stirred at 50 rpm paddle speed. Fig. 1 (right panel) illustrates ITZ release from as-received ITZ crystalline powder, aqueous spray-dried 1:5 ITZ powder:Sol without wet-media milling (W-AR-Sol-1:5) and with aqueous wet media milling prior to spray drying (W-Sol-1:5 and W-Sol-1:5-SDS, latter including SDS), a traditional nanocomposite prepared with 5:1 mass ratio of ITZ:Sol (W-Sol-5:1), and the spray dried solution of 1:5 ITZ-Sol in dichloromethane (S-Sol-1:5). Considering the ITZ solubility as a baseline, we see from Fig. 1 that the spray drying of aqueous as-received ITZ microcrystals–Sol suspension (W-AR-Sol-1:5) slightly increased the dissolution rate and extent of relative supersaturation of ITZ. Despite having a much higher ITZ:Sol ratio (W-Sol-5:1), a traditional nanocomposite with wet-milled ITZ exhibited a higher supersaturation than W-AR-Sol-1:5. On the other hand, upon use of higher Sol in the wet-milled ITZ suspension (W-Sol:1:5), a HyNASD was formed, which exhibited a notably higher supersaturation. The addition of SDS (W-Sol:1:5-SDS) slightly helped to raise the supersaturation. The ASD (S-Sol-1:5) was superior to the identical HyNASD formulation (W-Sol:1:5) in terms of ITZ supersaturation.

Other results from our dissolution study can be summarized as follows: without SDS in the formulation, 360%, 320%, and 790% ITZ relative supersaturation was achieved from HPC, VA64, and Sol-based HyNASDs, respectively, in the dissolution tests whereas the presence of SDS resulted in 480%, 360%, and 850% supersaturation from HPC, VA64, and Sol-based HyNASDs, respectively. These level of supersaturation generation capability of HyNASDs could be attributed to the slow recrystallization of ITZ in the dissolution medium, interactions/miscibility of ITZ–polymers as well as the size of the drug (nano)crystals in the polymeric matrix. Higher supersaturation from Sol-based HyNASDs compared to HPC/VA64-based HyNASDs could be explained by the greater miscibility of Sol–ITZ and micellar solubilization of ITZ by Sol. Presence of SDS in the formulation enhanced the relative wetting effectiveness of the polymers and resulted higher extent of supersaturation during dissolution due to the finer ITZ particles in the milled suspensions and higher amorphous content in the spray-dried powders compared to the formulation w/o SDS.

Overall, the most striking finding from this study is that despite having ~70–80% nanocrystals, HyNASDs provided very high extent of ITZ supersaturation (~800% within 210 min) unlike traditional nanocomposites (50%), which could render nanoparticle formulations more attractive in bioavailability enhancement of poorly soluble drugs. On the other hand, ITZ release comparison of the ASD and the HyNASD with identical formulation suggests that the ASD generated higher extent of supersaturation (up to 1670%) than HyNASDs (up to 790%). However, this extent of high relative supersaturation from a largely nanocrystalline drug formulation is an interesting and exciting finding, which could render HyNASDs competitive to ASDs upon further formulation–process optimization.

References

1. Müller RH, Gohla S, Keck CM. State of the art of nanocrystals–special features, production, nanotoxicology aspects and intracellular delivery. Eur J Pharm Biopharm. 2011;78(1):1–9.

2. Brough C, Williams R. Amorphous solid dispersions and nano-crystal technologies for poorly water-soluble drug delivery. Int J Pharm. 2013;453(1):157–66.

3. Bhakay A, Rahman M, Dave RN, Bilgili E. Bioavailability enhancement of poorly water-soluble drugs

via nanocomposites: Formulation processing aspects and challenges. Pharmaceutics. 2018;10(3):86.

4. Junghanns J-UA, Müller RH. Nanocrystal technology, drug delivery and clinical applications. Int J Nanomed. 2008;3(3):295–309.

5. Müller R, Jacobs C, Kayser O. Nanosuspensions as particulate drug formulations in therapy: rationale for development and what we can expect for the future. Adv Drug Delivery Rev. 2001;47(1):3–19.

6. Hancock BC, Parks M. What is the true solubility advantage for amorphous pharmaceuticals? Pharm Res. 2000;17(4):397–404.

7. Baghel S, Cathcart H, O'Reilly NJ. Polymeric amorphous solid dispersions: a review of amorphization, crystallization, stabilization, solid-state characterization, and aqueous solubilization of biopharmaceutical classification system class II drugs. J Pharm Sci. 2016;105(9):2527–44.

8. Rahman M, Ahmad S, Tarabokija J, Bilgili E. Roles of surfactant and polymer in drug release from spray-dried hybrid nanocrystal-amorphous solid dispersions (HyNASDs). Powder Technol. 2020;361:663–78.

9. Baird JA, Van Eerdenbrugh B, Taylor LS. A classification system to assess the crystallization tendency of organic molecules from undercooled melts. J Pharm Sci. 2010;99(9):3787–806.

10. Rahman M, Arevalo F, Coelho A, Bilgili E. Hybrid nanocrystal–amorphous solid dispersions (HyNASDs) as alternative to ASDs for enhanced release of BCS Class II drugs. Eur J Pharm Biopharm. 2019;145:12–26.

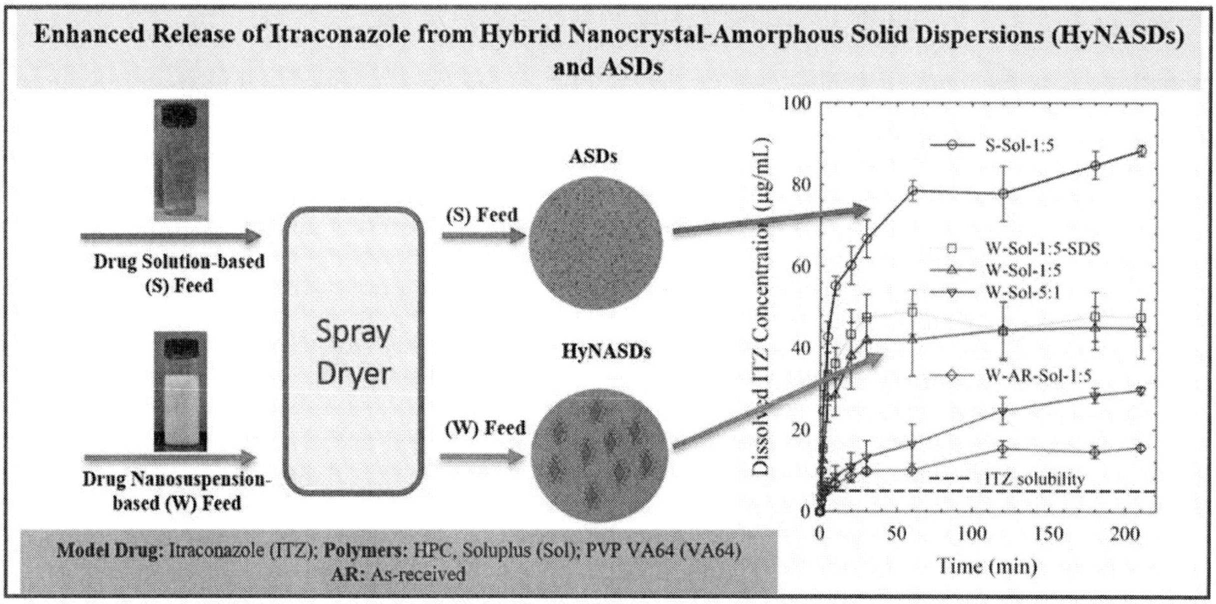

Fig. 1. Schematic of the preparation of ASDs and HyNASDs and their drug (ITZ) release during *in vitro* dissolution tests.

630044 Clarifying the Formulation Effect on the Drying Behavior of Pharmaceutical Granules after Wet Granulation

Abstract:

Pharmaceutical (wet) granulation aims to reduce the effect of the initial powder properties and replace them with the enhanced granule properties, this in order to enable downstream processing and tableting. The question remains as to which degree the initial properties of the active pharmaceutical ingredient (API) affect the process behavior in the subsequent fluid bed drying stage. This is important knowledge for the mechanistic modeling of this complex process: this constitutes whether any drying model should include the effect of API properties on the evaporation of the remaining water from the granules. This is a pivotal stepstone in the development of a generic drying model, which predicts drying behavior in function of formulation properties.

The goal of this work is therefore to quantify the influence of the API properties on the general drying behavior of granules, present in a fluid bed dryer of a continuous wet granulation line for pharmaceutical tablet production. This research was conducted using a ConsiGma-25™ semi-continuous fluid bed dryer (GEA), which is fed with wet granules from a twin-screw granulator through pneumatic transport in a tube. This dryer is characterized by six separate drying cells over which the continuous flow of wet granules is divided. A sequential filling pattern and parallel drying cycles allow this dryer to handle the continuous inflow. The key product, being the amount of granules that was loaded into a cell, subsequently is dried in this cell and discharged all at once downstream, where the continuous material stream is restored.

Recently, a new mechanistic multi-level model for this drying system was developed within our research group, starting from several previously described drying models. Each model describes the drying behavior on a different level. The population balance model (PBM) described by Peglow et al [1] was used as the base to describe the evolution of the drying air conditions in the drying cell such as the wall temperature, the air temperature, and the relative air humidity according to the bed height, assuming a bubbling fluidization regime. This is considered to be the bulk level of the model. This basic structure was then complemented with the single particle drying model described by Mortier et al [2] to obtain more accurate predictions of the granular drying behavior, such as the dynamics in granule temperature and the residual moisture content. This constitutes the drying behavior on a granular level. Besides that, this model was extended in order for it to simulate a continuous granule inflow and predict the drying behavior of wide granule size distributions. This multi-level drying model was validated for two different formulations from which it was concluded that the obtained model has a sufficient amount of predictive power for those APIs.

This study is the next building block for making the model generic for API properties, such as solubility, hydrophilicity, or grade. This thus leads to the API-level component in the modelling of drying behavior. To extend the existing mechanistic multi-level drying model, a relevance study on the influence of API properties on the overall drying behavior is performed. It is, however, important to make the trade-off between the additional required computing power and the increased accuracy of the predictions when an extra level is added to an existing model structure. In this experimental study, two different formulations with different API properties were granulated in such a way that comparable granules were obtained, considering the size distribution, the porosity, and the initial moisture content. Additionally, an extra API grade of one of the previously mentioned formulations was also included to account for API powder size. The dosages of the API were high, constituting 50 wt% of the formulation.

In total, six different drying experiments, supplemented by three repeated center point experiments, were performed on the similar granulates of each formulation under investigation. In this study, three different comparison points related to the granule properties were included. Complemented by varying the inlet air temperature between 40 and 60°C and the filling time, governing the cell load, between 90 and 180s. Each experimental point was conducted for, at least, three different drying times, ranging from 10 seconds added to the filling time to maximally two and a half times the filling time. The high-resolution PSD of the dried granules was collected using a particle size analyzer, QICPIC (Sympatec), with a dispensing system, Gradis (Sympatec). The residual moisture content is measured by a NIR-CI camera system, according to the method described in Ghijs et al. [3].

One of the most important conclusions following from the newly gathered data are the different breakage patterns, due to the fluidization process and/or pneumatic transport, for similar granules prepared with different formulations. This observation implies that the present granule size distribution, after certain drying times, differs according to the formulation properties. It is obvious that the granule size has a major influence on the evaporation rate and consequently the drying behavior.

From this perspective, it is useful to include this additional level, which considers the formulation properties, in the multi-level mechanistic fluid bed drying model. A comprehensive and accurate mechanistic drying model constitutes an important step towards generic models, that can be utilized for model-based process analysis or process control of continuous manufacturing in the pharmaceutical industry. More importantly, the generic model should be able to predict the drying behavior for new APIs where data on manufacturing is scarce. Such models are thus essential to facilitate the industry's transition to continuous manufacturing of pharmaceutical tablets and the effect of API in the drying behavior is an important factor in this.

References:

[1] Peglow, M., Kumar, J., Heinrich, S., Warnecke, G., Tsotsas, E., Mörl, L., & Wolf, B. (2007). A generic population balance model for simultaneous agglomeration and drying in fluidized beds. *Chemical Engineering Science,* 62(1-2), 513-532.

[2] Mortier, S. T. F. C., De Beer, T., Gernaey, K. V., Vercruysse, J., Fonteyne, M., Remon, J. P., Vervaet, C., Nopens, I. (2012). Mechanistic modelling of the drying behaviour of single pharmaceutical granules. *European Journal of Pharmaceutics and Biopharmaceutics*, Vol. 80(3), 682–689

[3] Ghijs, M., Vanbillemont, B., Nicolaï, N., De Beer, T., & Nopens, I. (2021). Two-dimensional moisture content and size measurement of pharmaceutical granules after fluid bed drying using near-infrared chemical imaging. *International journal of pharmaceutics,* 595.

628498 Assessing the Impact of API Surface Engineering on Drug Content Uniformity of Low-Dose Tablet Blends

Abstract:

Tablets having wide ranges of dosages are the preferred form of solid dosage form[1,2]. However, for pediatric or highly potent tablets, the major challenge is to achieve uniformity of active pharmaceutical ingredients (API)[1,3]. For that purpose, the APIs are often micronized to improve the drug content uniformity, where micronization is also expected to promote faster dissolution and subsequent absorption[3,4]. However, once the powders are micronized, their cohesiveness increases[5,6] forming agglomerates and thus adversely impacting the content uniformity of the powder blends, not to mention, often requiring wet granulation[3]. The recent study by Huang et al. 2017 demonstrated that the dry coating technique could alleviate the problem while avoiding wet granulation since it can significantly decrease interparticle cohesion[7], hence mitigating the adverse impacts from typical large agglomerates[3]. Although that work demonstrated the benefit of the dry coating technique, the model API was not poorly water-soluble, and the formulation consisted of a drug-excipient binary mixture. However, most tablet formulations consist of multi-component excipient powders and increasingly involve poorly water-soluble APIs. Therefore, the current work addresses that gap by investigating the effect of dry coated poorly water-soluble API (ibuprofen, d_{50} of 10 to 13 microns) within the multi-component tablet blends having 4 additional components that are different types of excipients: lactose monohydrate, a filler, micro cellulose crystalline, a tablet binder and a filler, crospovidone, a disintegrant, and magnesium stearate, a lubricant during tableting operations. For eliminating the confounding effect from size driven segregation[8] during mixing, all these components were selected to be in similar size ranges as the API[4]; i.e., crospovidone, magnesium stearate, lactose monohydrate, micro cellulose crystalline have mean particle size (d_{50}) of 38μm, 8μm, 19μm, and 20 μm, respectively. The changes in the bulk powder properties (flowability, bulk, and true densities) and tablet properties (API release rate, and tensile strength) were also measured and analyzed. In addition, the use of two different types of silica for dry coating were considered, following a recent work (Kim et al. 2021[9]), which demonstrated complex combined effect from agglomerate size reduction and surface wettability, both induced by the silica type, amount, and coating effectiveness. The experimental results demonstrated that even with a 5-components powder blend system, the dry coating of the micronized and cohesive poorly water-soluble API significantly improved the content uniformity of API at each of three, 1, 3, and 5%, drug loadings. Considering there are rather low drug loadings, improvements in powder flowability or bulk densities after dry coating were minimal as was expected. Interestingly, there were noticeable changes in the tablet tensile strength and API release rates when the tablets contain dry coated API. The relationship between the surface energy of the dry coated API and the API release rate from the multi-component tablets was also examined.

References

1. Lopez, F. L.; Ernest, T. B.; Tuleu, C.; Gul, M. O., Formulation approaches to pediatric oral drug delivery: Benefits and limitations of current platforms. *Expert Opinion on Drug Delivery* **2015,** *12* (11), 1727-1740.

2. Bhutani, U.; Basu, T.; Majumdar, S., Oral Drug Delivery: Conventional to Long Acting New-Age Designs. *European Journal of Pharmaceutics and Biopharmaceutics* **2021,** *162*, 23-42.

3. Huang, Z.; Xiong, W.; Kunnath, K.; Bhaumik, S.; Davé, R. N., Improving blend content uniformity via dry particle coating of micronized drug powders. *European Journal of Pharmaceutical Sciences* **2017,** *104*, 344-355.

4. Kunnath, K.; Huang, Z.; Chen, L.; Zheng, K.; Davé, R., Improved properties of fine active pharmaceutical ingredient powder blends and tablets at high drug loading via dry particle coating. *International Journal of Pharmaceutics* **2018,** *543* (1-2), 288-299.

5. Nase, S. T.; Vargas, W. L.; Abatan, A. A.; McCarthy, J. J., Discrete characterization tools for cohesive granular material. *Powder Technology* **2001,** *116* (2-3), 214-223.

6. Castellanos, A.; Valverde, J. M.; Quintanilla, M. A. S., Fine cohesive powders in rotating drums: Transition from rigid-plastic flow to gas-fluidized regime. *Physical Review E - Statistical Physics, Plasmas, Fluids, and Related Interdisciplinary Topics* **2002,** *65* (6).

7. Chen, Y.; Yang, J.; Dave, R. N.; Pfeffer, R., Fluidization of coated group C powders. *AIChE Journal* **2008,** *54* (1), 104-121.

8. Figueroa, I.; McCarthy, J. J., Using janus particles to control mixing and segregation of adhesive particle systems. *Industrial and Engineering Chemistry Research* **2010,** *49* (11), 5204-5214.

9. Kim, S.; Biligili, E.; Dave, R. N., Impact of hydrophobicity of dry coating material on the dissolution of micronized BCS class II drug powders. *Submitted to Interntional Journal of Pharmaceutics* **2021**.

627730 A Combined Kinetic–Microhydrodynamic Analysis of the Fenofibrate Nanosuspension Production in a Wet Stirred Media Mill

Abstract:

Majority of drug molecules coming out of high-throughput screening have poor water solubility due to their high lipophilicity and molecular weight [1]. 40% of the marketed drugs are poorly water-soluble, which results in low bioavailability. There are several formulation approaches to overcome this challenge such as amorphous solid dispersions, lipid-based formulations, and nanoparticle-based formulations [2]. Drug nanoparticles have been commonly used as they are shown to improve the dissolution rate and bioavailability of drugs due to increased specific surface area, higher saturation solubility as well as a reduction in the thickness of the diffusion layer [3]. These advantages led to the use of nanosuspensions or nanocomposites in several marketed products that have various delivery routes such as oral, parental, pulmonary, ocular, and dermal [4].

In the preparation of drug nanosuspensions, wet stirred media milling (WSMM) appeared to be the most preferred process due to its several advantages, according to a recent literature survey [5]. WSMM has the capability of producing high drug-loaded, stable nanosuspensions and it is a robust, reproducible, scalable, solvent-free, and environmentally benign process [6]. Despite its significant advantages, WSMM has also serious challenges. Most studies on WSMM focus on formulation challenges such as aggregation and crystal growth via Ostwald ripening (see [5] and references cited therein). In contrast, relatively scant information is available about the processing–operational challenges such as energy-intensive operation, high cost due to high energy consumption, long operating hours, and contamination of drug particles by the beads [6,7]. Overcoming these challenges entails a mechanistic understanding of the impact of process variables such as the stirrer speed, the size–material of the beads, and the bead loading on the breakage kinetics, milling time required for the desired product fineness, specific energy consumption, and media wear [8,9].

This study aims to examine the impact of stirrer speed ω, volumetric bead loading c, and bead material type (CPS: cross-linked polystyrene and YSZ: yttrium-stabilized zirconia) on the fenofibrate (FNB) particle breakage during WSMM via three breakage kinetic models and a microhydrodynamic model. FNB was used as a challenging, model poorly water-soluble drug because it is highly prone to aggregation and Ostwald ripening [10,11], unlike griseofulvin used in the previous microhydrodynamic studies [8,12,13] and its milling is relatively slow. In the experiments, 235 g suspensions with 10% FNB, 7.5% HPC-L, and 0.05% SDS were prepared, wherein HPC-L and SDS served as the stabilizers selected based on our prior work [10,11]. The evolution of FNB median particle size was tracked via laser diffraction during WSMM operating at 3000–4000 rpm with 35–50% (v/v) concentration of polystyrene and zirconia beads. First-order, n^{th}-order, and warped-time breakage kinetic models were fitted to the kinetic data with the objective of identifying the best kinetic model based on statistical analysis and physical plausibility considerations. The evaluation of breakage kinetic parameters with respect to the low and high values of the operating parameters was made via main effects plots. Main effects plots helped to visualize the influence of the milling variables on the breakage kinetics and microhydrodynamic parameters. Based on a recent modification [14] of the microhydrodynamic theory [15], several microhydrodynamic parameters, i.e., granular temperature θ, average bead oscillation velocity u_b, frequency of a single bead oscillation n, maximum contact pressure σ_b^{max}, radius of bead contact circle α_b, average frequency of drug particle compressions a, and pseudo energy dissipation rate for the drug particles $\Pi\sigma_y$ were calculated using the measured process variables, power consumption, and suspension viscosity–density. A subset selection algorithm was used along with Multiple Linear Regression Model (MLRM) to delineate how the breakage rate constant is affected by the microhydrodynamic parameters.

When the fitting capability of the first-order, n^{th}-order and warped-time kinetic models is compared (see Fig. 1) and statistically analyzed, n^{th}-order and warped-time kinetic models were found to better fit the experimental data than the first-order kinetic model. We assessed n^{th}-order and warped time models via main effects plots with changing process conditions and statistical analysis. The warped-time model was found hard to interpret because both of its paramaters n and k_0 drastically changed in opposite directions with increasing intensity of the process conditions. On the other hand, the nth-order model with $n \cong 2$ turned out to be the best overall model that explained the temporal evolution of the median size well. Whereas the variations in n and limiting particle size $d_{\lim}$ upon process parameter changes were relatively small, k was the sole parameter that directly and significantly changed with the process parameters. Therefore, the nth-order model was used for correlating the breakage rate constant k to the physical microhydrodynamic parameters.

The microhydrodynamic parameters provided valuable insights and a physical basis for the observed breakage behaviors under different operating conditions An increase in the stirrer speed led to increase in all microhydrodynamic parameters, thus favouring faster breakage. While an increase in bead loading led to two counteracting effects, i.e., more frequent, albeit less energetic bead–bead collisions, faster breakage was observed at the higher bead loading. YSZ beads led to faster breakage than CPS beads, which can be attributed to the higher values of all microhydrodynamic parameters, except the radius of contact circle α_b. Finally, a correlation between the breakage rate constant k and 4 microhydrodynamic parameters, i.e., maximum contact pressure σ_b^{max}, radius of contact circle α_b, average frequency of drug particle compressions a, and pseudo energy dissipation rate for the drug particles $\Pi\sigma_y$ was established. A statistically significant (p-value ≤ 0.01) MLRM of three microhydrodynamic parameters explained the variation in the breakage rate constant k best ($R^2 \geq 0.99$). This optimal model demonstrated that the average frequency of drug particle compressions interacting with the pseudo energy dissipation rate and the radius of the bead–bead contact circle govern the breakage kinetics in WSMM.

Overall, this study has offered the first comprehensive treatment of breakage kinetics during WSMM in view of the fundamental physics (microhydrodynamics) and revealed the microhydrodynamic parameters that govern the breakage kinetics. In a future study, we will combine the microhydrodynamic model with a population balance model to predict the evolution of the whole particle size distribution.

REFERENCES

1. Ding Z, Wang L, Xing Y, Zhao Y, Wang Z, Han J. Enhanced Oral Bioavailability of Celecoxib Nanocrystalline Solid Dispersion based on Wet Media Milling Technique: Formulation, Optimization and In Vitro/In Vivo Evaluation. Pharmaceutics. 2019;11(7):328.

2. Junyaprasert VB, Morakul B. Nanocrystals for enhancement of oral bioavailability of poorly water-soluble drugs. Asian Journal of Pharmaceutical Sciences. 2015;10(1):13-23. doi: https://doi.org/10.1016/j.ajps.2014.08.005.

3. Nakach M, Authelin J-R, Perrin M-A, Lakkireddy HR. Comparison of high pressure homogenization and stirred bead milling for the production of nano-crystalline suspensions. Int J Pharm. 2018;547(1):61-71. doi: https://doi.org/10.1016/j.ijpharm.2018.05.042.

4. Malamatari M, Taylor KMG, Malamataris S, Douroumis D, Kachrimanis K. Pharmaceutical nanocrystals: production by wet milling and applications. Drug Discovery Today. 2018;23(3):534-47. doi: https://doi.org/10.1016/j.drudis.2018.01.016.

5. Bhakay A, Rahman M, Dave RN, Bilgili E. Bioavailability Enhancement of Poorly Water-Soluble Drugs via Nanocomposites: Formulation–Processing Aspects and Challenges. Pharmaceutics. 2018;10(3):86. doi: 10.3390/pharmaceutics10030086.

6. Li M, Azad M, Davé R, Bilgili E. Nanomilling of Drugs for Bioavailability Enhancement: A Holistic Formulation-Process Perspective. Pharmaceutics. 2016 doi: 10.3390/pharmaceutics8020017.

7. Bilgili E, Guner G. Mechanistic Modeling of Wet Stirred Media Milling for Production of Drug Nanosuspensions. AAPS PharmSciTech. 2020;22(1):2. doi: 10.1208/s12249-020-01876-w.

8. Afolabi A, Akinlabi O, Bilgili E. Impact of process parameters on the breakage kinetics of poorly water-soluble drugs during wet stirred media milling: A microhydrodynamic view. Eur J Pharm Sci. 2014;51:75-86. doi: https://doi.org/10.1016/j.ejps.2013.09.002.

9. Bilgili E, Capece M, Afolabi A. Modeling of milling processes via DEM, PBM, and microhydrodynamics. In: Pandey P, Bharadwaj R, editors. Predictive Modeling of Pharmaceutical Unit Operations. Woodhead Publishing; 2017. p. 159-203.

10. Azad M, Afolabi A, Bhakay A, Leonardi J, Davé R, Bilgili E. Enhanced physical stabilization of fenofibrate nanosuspensions via wet co-milling with a superdisintegrant and an adsorbing polymer. Eur J Pharm Biopharm. 2015;94:372-85. doi: https://doi.org/10.1016/j.ejpb.2015.05.028.

11. Knieke C, Azad MA, Davé RN, Bilgili E. A study of the physical stability of wet media-milled fenofibrate suspensions using dynamic equilibrium curves. Chem Eng Res Des. 2013;91(7):1245-58. doi: https://doi.org/10.1016/j.cherd.2013.02.008.

12. Parker N, Rahman M, Bilgili E. Impact of media material and process parameters on breakage kinetics–energy consumption during wet media milling of drugs. Eur J Pharm Biopharm. 2020;153:52-67. doi: https://doi.org/10.1016/j.ejpb.2020.05.024.

13. Li M, Alvarez P, Bilgili E. A microhydrodynamic rationale for selection of bead size in preparation of drug nanosuspensions via wet stirred media milling. Int J Pharm. 2017;524(1):178-92. doi: https://doi.org/10.1016/j.ijpharm.2017.04.001.

14. Guner G, Yilmaz D, Bilgili E. Breakage kinetic models and microhydrodynamic correlations for wet stirred media milling of fenofibrate. Drug Dev Ind Pharm. 2021, under review.

15. Eskin D, Zhupanska O, Hamey R, et al. Microhydrodynamic analysis of nanogrinding in stirred media mills. AIChE J. 2005;51(5):1346-1358, doi: https://doi.org/10.1002/aic.10392.

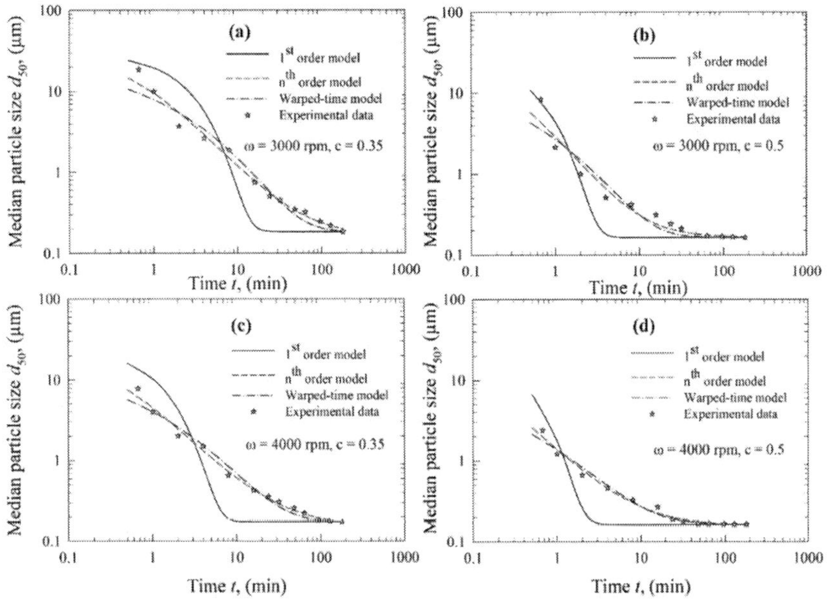

Fig. 1. Temporal evolution of the median particle size d_{50} during the wet milling of fenofibrate with CPS beads and fitting of the experimental data by various breakage kinetic models.

631287 Cross-Linked Biopolymer Nanosuspension Preparation: Impact of Solvents

Abstract:

Biopolymers (cellulose, starch, collagen, silk, chitosan, polylactic acid, etc.) nanoparticles have received considerable attention in diverse applications such as drug delivery, pharmaceuticals, biomedical, food, cosmetic, environmental, etc. [1–4]. Nano-sized or colloidal superdisintegrants, a subclass of cross-linked biopolymers, prepared by wet stirred media mill, have shown great promise as a functional additive for enhanced stabilization of drug nanosuspensions and fast drug nanoparticle recovery during redispersion and immediate drug release from nanocomposites during *in vitro* dissolution tests [2,3,5,6]. Wet stirred media milling (WSMM) is able to comminute materials into smaller sizes than most milling approaches [7], either ambient dry or cryogenic conditions. Recent studies reveal that WSMM has been the most widely used process for preparing nanosuspensions in the pharmaceutical field because WSMM is a robust, reproducible, and scalable process [8,9]. The number of stress events per unit time and unit volume and the stress intensity is very high in WSMM. Hence, stirred media mills' specific energy consumption for producing very fine particles is less than that of other mills [10]. Hence, stirred media mills' specific energy consumption for producing very fine particles is less than that of other mills [10]. Moreover, excellent turbulent mixing of the mill content and the presence of a liquid medium in the mill allows for fast convective heat transfer and cooling during milling, which minimizes the occurrence of hot spots and associated product degradation. Overall, the use of WSMM for the preparation of cross-linked biopolymer nanosuspensions is a very promising route, and outstanding contributions from various groups were reported [2,11,12]. In the current work, understanding the impact of solvent (aqueous vs. non-aqueous) and swelling on the particle size and breakage kinetics during WSMM in the presence/absence of stabilizing polymer–surfactant is addressed.

The impact of two solvents (water vs. acetone) in the preparation of cross-linked biopolymer nanosuspensions in WSMM was investigated. Sodium starch glycolate (SSG), a model cross-linked biopolymer, was selected owing to its popularity in the pharmaceutical industry [8]. SSG was milled in water (M1) and aqueous solutions of hydroxypropyl cellulose (HPC) and sodium dodecyl sulfate (SDS) as stabilizers (M2) as well as in acetone, with (M3) and without (M4) the same set of stabilizers. The use of HPC–SDS in M2 and M4 is expected to help prevent aggregation, depending on the liquid medium. The milling dynamics were investigated by studying the evolution of the particle size distribution (PSD) during milling via laser diffraction. It is hypothesized that the solvent affects particle breakage–aggregation, swelling, and microhydrodynamics in the mill and, in turn, PSD evolution. The impact of the solvents on the SSG swelling, suspension viscosity, and the microhydrodynamic parameters was examined. Detailed analysis of milling dynamics, swelling data, and SEM images are presented for the purpose of developing a better understanding of the roles played by the solvents.

The model cross-linked biopolymer, SSG, was milled for 4 h in DI water (swelling solvent) and acetone (non-swelling solvent) with and without stabilizers. A combination of HPC–SDS was used as a stabilizer system. Particle sizes of SSG particles in various states, i.e., as a dry-sieved powder (measured by Rodos/Hellos system), swollen in water (measured by Coulter LS 13 320), dispersed in acetone, and after 4 h milling in acetone and water is shown in Fig. 1a. Whereas the SSG particles in the water had a larger median size (114 mm) than the dry SSG particles (32 mm), there was no discernible size difference between the dry SSG particles and those dispersed in acetone. This observation can be explained by the fact that SSG does not swell in acetone, but it does so in water; SSG has a hydration capacity of 18.3 g water/g SSG [13], and we estimate that about 38% of water was absorbed by SSG. A cursory look at the particle size statistics in Fig. 1a pointed to a distinct size difference when SSG was milled in water vs. acetone for 4 h, signifying the significant impact of the solvents. Both 50% and 90% passing sizes of the cumulative PSD (d_{50} and d_{90}, respectively) of the swollen SSG particles were drastically reduced to the colloidal domain, with a slight positive effect of the stabilizers (M2 vs. M1). These suspensions are referred to as nanosuspensions in prevalent pharmaceutical terminology, although the median size was slightly above 100 nm. On the other hand, the dry (unswollen) SSG particles in acetone did not exhibit such a drastic extent of size reduction; coarse micron-sized suspensions with a wide PSD were formed. A comparison of M1 and M3 suspensions in Fig. 1b shows that no settling occurred with the nanoparticles in the well-dispersed M1 suspension after 3-day storage, whereas settling occurred in the coarse M3 suspension with micron-sized particles. A particle size difference, i.e., d_{50}: 4.2 vs. 2.7 mm between M3 and M4 suspensions indicates that stabilizers reduced the extent of particle aggregation in acetone.

In a short-term physical stability test, the PSD of milled suspensions after 3 days of storage indicates insignificant particle growth or aggregation in the aqueous suspensions; these suspensions were physically stable. On the other hand, the non-aqueous (acetone) suspensions had larger particles upon storage for 3 days, signifying continued aggregation during the storage. The stabilizers helped to slow down SSG aggregation in acetone, but they were unable to suppress the aggregation completely. The results suggest that SSG particles swelled in water and underwent extensive breakage into nanoparticles, and these fine particles did not aggregate even in the absence of HPC–SDS. The presence of HPC–SDS reduced the extent of aggregation and SSG particle size in acetone. The presence of larger primary SSG particles in the SEM images (not shown for brevity) of 4 h milled water vs. acetone suspension samples qualitatively suggests that both the rate and extent of particle breakage were higher in water than in acetone. Our microhydrodynamic analysis of the milling data suggests that while swelling-induced coarsening enhanced SSG particle capture probability and achieved faster breakage initially, acetone was generally more favorable for faster breakage of $< \sim 10$ μm particles than water. Despite this slight advantage of acetone, water was overall more favorable for faster production of stable nanosuspensions than acetone due to swelling-induced coarsening and softening of SSG particles as well as favorable nanoparticle stabilization in water.

We have demonstrated a significant influence of the solvents on the PSD evolution of cross-linked biopolymeric particles during WSMM through their modulation of the complex interplay between breakage–aggregation–stabilization, microhydrodynamics–rheology, and swelling. In a future study, the swelling-induced softening of SSG particles and adsorption of HPC on SSG particles in acetone should be studied to gain a deeper fundamental understanding. Moreover, a formulation optimization study may be performed to identify the optimal concentration of HPC–SDS to suppress SSG particle aggregation in acetone completely or find alternate molecular weight/type of soluble polymers to achieve the same.

References

[1] H.-Y. Kim, S.S. Park, S.-T. Lim, Preparation, characterization and utilization of starch nanoparticles, Colloids and Surfaces B: Biointerfaces, 126 (2015) 607-620.

[2] M. Azad, A. Afolabi, A. Bhakay, J. Leonardi, R. Davé, E. Bilgili, Enhanced physical stabilization of fenofibrate nanosuspensions via wet co-milling with a superdisintegrant and an adsorbing polymer, European Journal of Pharmaceutics and Biopharmaceutics, 94 (2015) 372-385.

[3] M. Azad, C. Arteaga, B. Abdelmalek, R. Davé, E. Bilgili, Spray drying of drug-swellable dispersant suspensions for preparation of fast-dissolving, high drug-loaded, surfactant-free nanocomposites, Drug Development and Industrial Pharmacy, 41 (2015) 1617-1631.

[4] A. Bhakay, M. Azad, E. Vizzotti, R.N. Dave, E. Bilgili, Enhanced recovery and dissolution of griseofulvin nanoparticles from surfactant-free nanocomposite microparticles incorporating wet-milled swellable dispersants, Drug Development and Industrial Pharmacy, 40 (2014) 1509-1522.

[5] E. Bilgili, R. Dave, A. Bhakay, M. Azad, Methods for superdisintegrant-based composite particles for dispersion and dissolution of active pharmaceutical agents, in, US Patent 2018, pp. 41 Pages.

[6] J. Guo, X. Gu, Y. Mai, Y. Zhao, G. Gou, J. Yang, Preparation and characterisation of tetrandrine nanosuspensions and in vitro estimate antitumour activity on A549 lung cancer cell line, Journal of Microencapsulation, 37 (2020).

[7] G.C. Lowrison, Crushing and Grinding: The Size Reduction of Solid Materials, Butterworths, 1974.

[8] A. Bhakay, M. Rahman, R.N. Dave, E. Bilgili, Bioavailability enhancement of poorly water-soluble drugs via nanocomposites: formulation‑processing aspects and challenges, Pharmaceutics, 10 (2018).

[9] M. Li, M. Azad, R. Davé, E. Bilgili, Nanomilling of drugs for bioavailability enhancement: A holistic formulation-process perspective, Pharmaceutics, 8 (2016).

[10] A. Kwade, Wet comminution in stirred media mills - research and its practical application, Powder technology., 105 (2000) 14.

[11] M.A. Azad, A. Afolabi, N. Patel, R. Dave, E. Bilgili, Preparation of stable colloidal suspensions of superdisintegrants via wet stirred media milling, Particuology, 14 (2014) 76-82.

[12] T. Fu, X. Gu, Q. Liu, X. Peng, J. Yang, Study on the stabilization mechanisms of wet-milled cepharanthine nanosuspensions using systematical characterization, Drug Development and Industrial Pharmacy, 46 (2020) 200-208.

[13] A. Quadir, K. Kolter, A Comparative Study of Current Superdisintegrants, Pharmaceutical Technology, (2006) S38-S42.

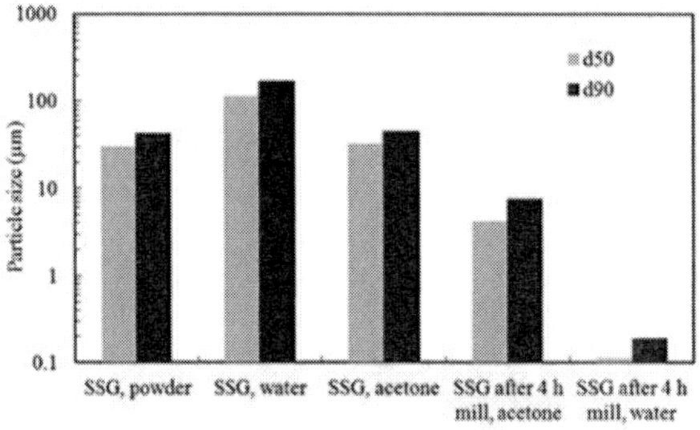

(a)

(b)

Fig. 1.Particle sizes of the SSG particles: (a) dry sieved particles, sieved particles swollen in water, particles in acetone, and 4 h milled particles in acetone (M3) and water (M1) without stabilizers; (b) photograph of the 4 h milled SSG suspensions with water (M1) and acetone (M3) (without stabilizers) after 3-day storage.

630324 Wurster Coating: Prediction of Undesired Agglomeration

Abstract:

Wurster fluid bed coating is an effective way of manufacturing pharmaceutical pellets requiring specific characteristics such as high drug loading or enteric protection. Optimization of the coating process can be demanding because it necessitates many time-consuming pilot-scale experiments, the scope of which is limited by the cost of active pharmaceutic ingredients. To reduce the need for experiments and prevent undesired agglomeration, a multi-scale model has been developed.
The model utilizes statistically evaluated particle dynamics as well as heat and mass transfer resolved for both device and particles. The model was validated against existing production data, and the predicted agglomeration probability was in agreement with the observed undesired agglomeration or successful coating regime.
Wurster fluid bed coating is one of the most versatile coating methods [1]. It offers fast heat and mass transfer, as well as temperature homogeneity. In contrast to conventional fluidized bed coating, a coating solution is sprayed from a nozzle placed bottom-up in the center of a cylindrical tube (draft tube). Particles are driven through the tube and periodically move through the spraying zone, where they collide with small droplets of the coating solution. The circulation of the particles in this way ensures good mixing and provides better control of particle movement and the resultant coating properties.
In comparison to other coating methods, the main disadvantage of Wurster is the tedious setup and process optimization [1]. Currently, manufacturers choose this method mainly for special applications like formulation of sustained or enteric release. Coloring, odor, or taste masking, where uniform film thickness and coating porosity are less important, can be done in a conventional top-spray fluidized bed.
In this work, we mainly focus on the coating of small particles for a long period, which is the usual case for the deposition of an API. The long process time assures the low inter-particle coating variability [2].
However, agglomeration is here the main concern. Agglomerated pellets must be separated before capsule filling, reducing the yield. If agglomerated pellets again disjoin before or during separation, the resulting inter-particle variability (API dosage) will be higher. If combined with sustained release or enteric protection, the whole formulation might be ineffective.
Agglomeration has several causes. An insufficient drying rate can lead to excessive wetness and result in the coating solution acting as a binder. Conversely, excessive drying or low air humidity can lead dry particles to generate electrostatic charges and, thereby, stick together. To prevent agglomeration, it is necessary to keep the wetness of the particles within a narrow threshold range while maintaining homogeneous coating distribution. However, the coating distribution and particle wetness can only be controlled indirectly by setting operating parameters, such as temperature, spraying rate, draft tube position, volumetric airflow rate, and humidity.
Agglomeration can be viewed in terms of probability as a product of geometrical and the physical collision success factors [3]. The geometrical factor is based on surface coverage, and the physical factor is based on coalescence conditions derived from Stokes number analysis [4].
CFD-DEM simulation can accurately predict the particulate flow behavior. Using CFD-DEM directly coupled with droplet deposition, drying, and particle liquid bridges interactions is so strongly limited by the number of particles that any application to a real pharmaceutical process where several millions of particles are present seems impossible. However, a promising approach is a statistical evaluation of the repetitive patterns in particle circulation [5].

We follow this idea and present a model that uses statistically evaluated particle dynamics. Additionally, at the onset of the agglomeration, the inlet spray rate, air humidity, and inlet air temperature have almost no impact on the particle dynamics. And the drying kinetics can be analyzed separately with lower computational effort. On the device-scale, the model predicts the temperature, moisture content, and average coating thickness. On the particle-scale, model computes particle dynamics via CFD-DEM. On the micro-scale, the model estimates the coating deposition, drying kinetics and change of its physical properties. By coupling models at all length-scales, including only statistical analysis of particle movement and Stokes coalescence condition, it is possible to evaluate particle agglomeration probability and use it as a tool for process optimization and possibly for transfer between different production sites and equipment.
We have analyzed data from 13 production records. All experiments were carried out in Glatt GPCG 2 fluid bed equipped with a 7" conical Wurster chamber and a cylindrical expansion chamber. Spraying solution consisted of ethanol or isopropanol together with a full pharmaceutical formulation including the API and polymer. And the predicted agglomeration probability was in agreement with the observed undesired agglomeration or successful coating regime.

[1] P. Hadfield, Processing and Equipment Considerations for Aqueous Coatings, (4th ed.), CRC Press, New York (2016), pp. 49-82

[2] X.X. Cheng, R. Turton, The prediction of variability occurring in fluidized bed coating equipment. II. The role of nonuniform particle coverage as particles pass through the spray zone, Pharm. Dev. Technol., 5 (2000), pp. 323-332

[3] P. Rajniak, F. Stepanek, K. Dhanasekharan, R. Fan, C. Mancinelli, R. Chern, A combined experimental and computational study of wet granulation in a Wurster fluid bed granulator, Powder Technol., 189 (2009), pp. 190-201

[4] L.X. Liu, J.D. Litster, S.M. Iveson, B.J. Ennis, Coalescence of deformable granules in wet granulation processes, AICHE J., 46 (2000), pp. 529-539

[5] T. Lichtenegger, S. Pirker, Recurrence CFD – a novel approach to simulate multiphase flows with strongly separated time scales, Chem. Eng. Sci., 153 (2016), pp. 394-410

622354 Accelerated Quantitative DEM Simulation of Large Screw Feeder and Tablet Press Feeder Using Adaptive Coarse Graining Technique

Abstract:

Coarse-graining of particles is unavoidable for practical DEM simulations. For pharmaceutical industry, the typical dimensions of the production scale equipment are ~ 1 m whereas the typical powder contains particles around 0.1 mm. This size ratio requires the coarse-graining ratio to be very high which introduces large inaccuracies in the simulation. To tackle this situation, recently an adaptive coarse-graining strategy has been proposed [1]. This newly proposed technique has been demonstrated using a simple hopper discharge. However, refinement of the algorithm and application for a variety of systems is pending. In this work we will apply the adaptive coarse-graining technique for two problems of practical importance: tablet die feeder and hopper screw feeder.

It will be shown that the adaptive coarse-grained simulation of tablet die filling reveals additional features that were not otherwise seen in the fixed coarse-grained simulations. This is because the model particles from fixed coarse-graining are much larger than actual particles that introduces artifacts near the die mouth (like exaggerated weight RSD and porosity).Jamming in the feeder section is also observed.As shown in figure 1, resolved particles can be taken near the mouth of the die for accurate estimation of the filling process, when adaptive coarse-graining technique is used. Slightly coarse-grained particles will be taken near the paddle wheel for correct estimation of the flow, and larger coarse-grained particles will be taken to fill up the feed chute. With this strategy, the system can be simulated accurately with reasonable computation.

A different type of challenge exists in implementing adaptive coarsegraining for the hopper screw feeder system. Because of the large recirculation near the feed end of the screw, adaptive coarse-graining induces repeated resolution and consolidation of same set of particles which eats up computational resources and renders adaptive coarse-graining less attractive for such systems. It will be shown that an innovative zoning strategy (snapshot given in figure 1) can resolve this issue and the adaptive coarse-graining can improve the efficiency of such systems substantially.

References:

1. A Multi-Level Coarse Graining Technique for Discrete Element Method (DEM) Simulation. Jayanta Chakraborty, Tarun De, Jitendra Kumar, Salvador García-Muñoz, Maitraye Sen. AIChE Annual Meeting, 2020

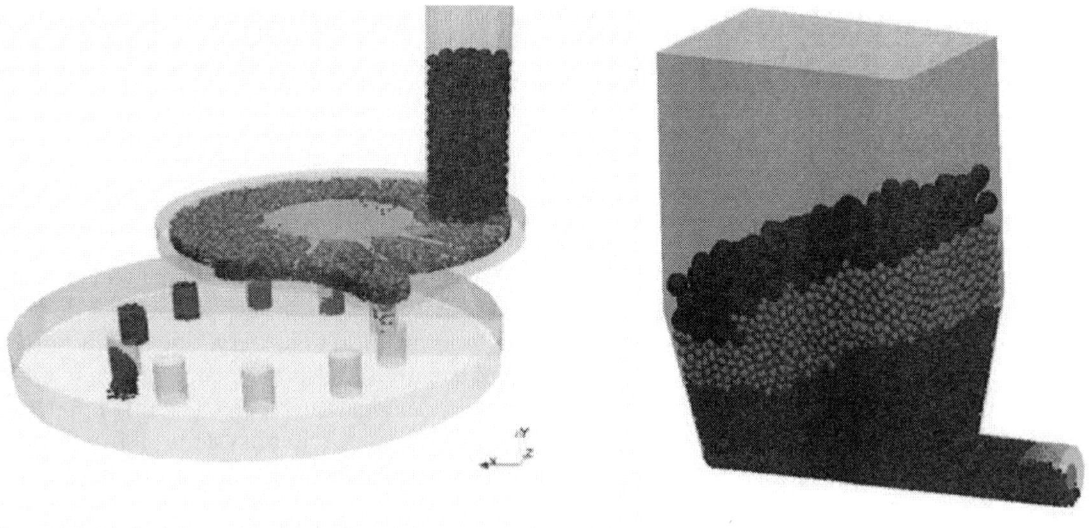

Figure 1: Snapshots of the adaptive coarse-grained simulation of tablet die feeder (left) and the screw feeder (right).

625348 Numerical Workflow for Predicting Tablet Coating Uniformity in Pharmaceutical Coating Processes from Lab to Commercial Scale

Abstract:

In this presentation, tablet coating uniformity was predicted in different coater geometries (ProCepT, Glatt GMPC1, Omega and Bohle BFC 5) using commercial software (Rocky DEM and ANSYS CFD). Two tablet shapes () and various operating conditions (drum rotational speed, batch load, spray rate, and nozzle position) were investigated and the results were compared against experimental data. Simulation results showed good agreement with both experimental and analytical models of the coating process, and a sensitivity study was also performed out to determine the influence of the critical parameters during the process. A workflow for predicting the Coefficient of Variation of the tablet film coating thickness using numerical simulation is proposed for virtual process design.

625580 Practical Applications of Dry-Coating for Pharmaceutical Tablet Manufacturing

Abstract:

Dry-coating processes designed to coat API or excipients with glidant (nano-sized silicon dioxide), have been shown effective to improve the manufacturability of formulations containing APIs with poor flowability and tabletability. Despite the advantages of dry-coating at the lab-scale using a wide variety of equipment such as powder blenders/mixers and mills, manufacturing-scale processes have yet to gain significant acceptance. This may be due to the poor scalability of dry-coating processes or presumably due to the aversion of risk associated with adopting new technology in the pharmaceutical industry. To address this issue, this work details the industrial application of dry-coating for use in tablet manufacturing.

In this study, enhancement to powder flowability and tabletability are discussed. Effects and issues related to dissolution as well as sticking/picking during tabletting are also presented. As a major novelty, scale-up studies utilizing a conical screen mill (comill) are detailed. While the comill is frequently employed in the production of tablets as a screening/de-lumping or particle size reduction process, it has not gained acceptance as a dry-coating process. This work shows that the comill can be used as an effective and efficient process for the purpose of dry coating at both lab and manufacturing scale. While our previous work has shown the effectiveness of the comill at the lab scale [1], this work will focus on matters of scale up. The parameters affecting development of the comill as a dry-coating process are examined.

[1] M. Capece, C. Borchardt, A. Jayaraman, Improving the effectiveness of the Comil as a dry-coating process: Enabling direct compaction for high drug loading formulations. Powder Technology 379 (2021) 617-629.

AbbVie funded and participated in the study design, research, data collection, analysis and interpretation of data, as well as writing, reviewing, and approving the publication. Maxx Capece and Jeffery Larson are AbbVie employees and may own AbbVie stock/options.

622638 Scroll Compaction: A Novel Approach for Uniform Compaction of Powder Beds

Abstract:

Consolidation / compaction of a powder bed is a common occurrence in powder handling processes. In many cases, this is intentional, for example during tabletting or isostatic pressing. However, in other scenarios consolidation of a powder is undesirable, leading to poor process performance. Consolidation reduces inter-particular spacing and increases the specific contact area between particles, increasing particle-particle interactions which in turn reduce powder flowability, and in some cases may prevent it altogether, especially when other factors such as environmental conditions are also in effect.

As such, understanding the impact of consolidation is a useful tool for identifying potential processing issues, even where consolidation is not a deliberate process step. Consolidation can occur via many different routes, the most common are vibrational forces, for example during transport, and compaction through the application of applied load, for example during storage where powders may compact under their own mass.

Commonly applied compressibility tests typically use single end compaction to consolidate the powder bed, i.e. the normal stress is applied on a single principal plane. However, consolidation is not homogeneously induced throughout the powder bed using this method, especially with highly compressible materials. One alternative method for preparing a more uniformly compacted powder is to build-up the powder bed layer by layer but this has practical limitations and can be extremely time consuming.

Scroll Compaction is a proposed solution. This approach uses a novel helical blade which rotates upwards through the powder bed, compacting at predefined intervals. Using this method, the operator is only required to fill the test vessel with sufficient material and start the program, no further input is required. This method is also quick; 15-20 compaction steps can be completed in approximately 5 minutes.

To demonstrate the capability of this novel technique, Scroll Compaction was compared with single stage compaction for several materials of varying degrees of cohesion. In order to determine if Scroll Compaction was a more effective method of consolidating materials, the resistance to flow of the compacted bed, presented as a flow energy, was measured alongside the bulk density. Increases in flow energy compared to that of the non-consolidated "as received" sample are indicative of a consolidated powder bed.

The results demonstrate that Scroll Compaction resulted in greater increases in both bulk density and flow energy values, compared with single stage compaction. The effect was more pronounced for cohesive materials, which tend to be less efficient at transmitting forces through the powder bed, however there were still clear differences between the two methodologies for free-flowing, typically non-compressible materials. Overall, the results suggest that Scroll Compaction is more effective for consolidating powders, resulting in a more uniformly, more process-relevant compacted bed.

623555 A DEM-Based Approach for Prediction of Pellet Mixing in a Full-Scale Conical Blender

Abstract:

Blending operations are prevalent in the manufacture of drug products within the pharmaceutical industry, but most often involve powders in diffusion/tumble blenders. In this work, the development of blending process for millimeter-scale pellets in a large-scale, conical screw mixer is discussed. A model-based approach using the discrete element method was used to guide process development. The DEM-based approach was used (1) predict the mixing time required for the full-scale process in a 6000 L conical mixer and (2) guide appropriate selection of blend uniformity acceptance criteria for pellet samples. An experimental mixing study at scale confirmed the mixing conditions and mixing time required to achieve a well-mixed batch.

WK, DM, MD, and BR are employees of AbbVie and may own AbbVie stock. DA is a former employee of AbbVie and may own AbbVie stock. AbbVie sponsored and funded the study; contributed to the design; participated in the collection, analysis, and interpretation of data, and in writing, reviewing, and approval of the final publication.

620597 Using Powder Rheometry to Identify and Quantify Different Humidity Induced Caking Behaviours

Abstract:

"Caking" is the phenomenon where materials gain strength during storage due to changes in particle-particle interactions. Whilst caking occurs via several different mechanisms, for example, mechanical, thermal, environmental and/or chemical, the absorption and migration of moisture is frequently the most dominant mechanism. Fully understanding the propensity to cake is important for minimising downstream process issues, such as blockages, which may have a detrimental impact on the productivity and final product quantity.

It is often assumed that caking occurs uniformly throughout a powder bed, however recent advances in powder rheometry techniques have demonstrated that powders can exhibit a wide range of caking behaviour, including non-homogenous caking (crusting) and even a "hybrid" response dependent on storage conditions. Furthermore, as powder rheometry can quantify the resistance to flow throughout the powder bed, progression of the cake/crust and its resultant strength can be assessed using this method.

To demonstrate the ability of powder rheometry to capture the impact of different caking behaviours, several materials including skimmed milk powder (SMP), sulphated methyl ester (SME) and three food flavours were stored for several days under controlled humidity conditions. SMP exhibited non-homogeneous behaviour with a clearly defined caked region which progressed through the sample over several days. In contrast SME exhibited the aforementioned hybrid behaviour, with the bulk of caking localised, but further caking occurring below the defined crust. In both cases, the strength of the crust was shown to increases over time and demonstrating that caking may require several days to fully develop. Analysis of the food flavours indicated that some materials can also switch between the observed caking regimes, with one material initially exhibiting hybrid behaviour before transitioning to a more homogeneous response.

Caking behaviour may be further complicated by variations in the environmental conditions. To further understand this, samples were stored at 75%RH for 3 days and followed by 3 days at ambient (33%RH). As expected, the return to ambient conditions decreased the degree of caking for some samples. However, for SMP, drying the sample had limited impact on the crusted region but more interestingly, caking continued to occur in the previously uncaked region below the crust. More notably one of the food flavourings presented a significant increase in the degree of caking on return to ambient conditions, demonstrating that for some materials, drying can actually exacerbate caking behaviour.

Overall, the results demonstrated the humidity induced caking is complex, manifesting in a diverse range of different behaviours. In order to fully understand the caking behaviour of a given material, it is therefore necessary to use a technique capable of capturing variations in the flowability throughout the powder bed. Though the use of powder rheometry it is not just possible to identify the caking regime but also quantify the impact, allowing the progression of the cake over an extended period be determined.

629290 Influence of Temperature on the Packing Dynamics of Powders

Abstract:

Numerous industrial processes involve granular materials and fine powders. However, due to the lack of complete theoretical descriptions, the understanding of powder behavior has to rely on the characterization of the macroscopic properties (flow, packing dynamics, agglomeration, segregation, ...). Powder behavior is known to be influenced by particles shape and size distributions, friction forces and cohesive interactions. Cohesive interactions are due to the presence of liquid bridges, electrostatic charges, van der Waals interactions or more rarely magnetic dipole-dipole interactions. For powders, the behavior is usually driven by these cohesive interactions due to the small particle sizes.

Environmental conditions (relative humidity and temperature) can alter the powder properties changing the predominance of these grain interactions. Indeed, moisture is known to influence both surface grains conductivity and capillary bridges formation with as a consequence a modification of the flowing properties. Furthermore, empirical observation seems to indicate that the interactions between the grains as well as grains breakage behavior are affected by temperature. However, precisely controlling the powder temperature during a process requires costly equipment and is almost not feasible for large production facilities. Seasonal changes and production in different sites over the globe induce unavoïdable variations of the environmental conditions at which the powder is processed. Therefore, understanding the influence of temperature on the powder behavior is of great interest to predict the degradation of process performance and establish optimal processing conditions. Investigations on the effect of environmental conditions as mainly focused on relative humidity [1-2], temperature has only be considered in processes that involve powder heating at high temperature, such as polymer and metal powders in additive manufacturing [3].

In the present study, we emphasized the importance of even low variations of temperature on common food and lactose powders. Changes in powder properties have been assessed by a packing dynamics characterization method which provides useful information on the powder properties that can be related to its flowability. An improved version of the classical tapped density method has been used (GranuPack, GranuTools)[4] and modified to allow continuous powder heating during the measurement. After each tap, the density is measured accurately with the use of an inductive sensor, allowing to extract useful information on the packing kinematics. Temperature is found to slow down the packing dynamics at temperature variations that are usually considered as normal room fluctuations (20 – 40 °C). The importance of evaluating the effect of temperature even for processes that do not involve powder heating is thus highlighted. We also demonstrate the inability of the classically used Hausner ratio analysis to properly describe this behavior. These new results evidence the need to take into account the environmental conditions at which the powder will be processed at the beginning of the development and characterization stage.

629369 Investigation of Electrostatic Properties of Lactose Powders: Implications for Dry Powder Inhaler (DPI) Processes

Abstract:

Dry powder inhaler (DPI) products are designed to deliver a dry powder active ingredient to a patient's lungs using an inhaler device. The formulation typically consists of a homogenous blend of the active ingredient and an excipient (typically lactose), which is contained within a reservoir, capsule, or blister. Such formulations are produced by physically mixing micronised drug particles and the larger-sized excipient 'carrier' particles. Contrary to many conventional tableting formulations, the excipient is an important functional ingredient in a DPI formulation. The excipient carrier plays at least two roles in the formulation. Since the active ingredient is used in very small quantities, blending with the excipient is necessary to enable delivery of such a small quantity. Second, the excipient must separate from the drug during inhalation so that the drug can be inhaled into the patient's lungs.

The excipient physical properties and interfacial forces affect the blend uniformity and drug release of the DPI formulation. Some of these properties include particle size distribution, flowability, surface chemistry, morphology, and electrostatic properties of the excipient. It is important that these characteristics are controlled during manufacture and storage of the excipient. The main challenges in DPI development are producing a uniform blend for accurate dosing and then proper drug separation during inhalation. Beyond that, reducing batch to batch variability is extremely challenging in these types of formulations.

Electrostatic charges of the excipient and drug particles (API) can affect the performance of dry powder inhalation drug formulations. Electrostatic forces on the excipient could cause the API particle to adhere to the surface of the excipient and prevent release. Also, electrostatic forces in the final blend could affect the powders release from the device. Of course, variability in charge profiles could cause variability in the final drug fine particle fraction (API that would be released to the patient). Characterization of these charges is an important tool to possibly allow a higher percentage of the API to be released from the excipient/device. Managing or controlling these charges could also be important to reduce the variability of API release. However, due to the complexity of the mechanisms at the origin of triboelectric effects, electrostatics has been only recently investigated for pharmaceutical powders [1,2].

In this study, the electrostatic properties of lactose excipients designed for DPI applications (AeroFlo, Kerry) have been investigated with the GranuCharge (GranuTools)[3]. Anhydrous and monohydrate powders differing from their particle size distribution have been selected. Moisture is controlled by conditioning the powders before the measurements at a fixed relative humidity with a specifically designed rotating tumbler coupled with a humid air generator (GranuMidity, Granutools). Then, powder charge density is measured before and after the powder flow through a Stainless Steel SS316L pipe. During the flow, the particle/particle and particle/pipe interactions lead to a charge buildup due to triboelectric effect, increasing the overall charge density of the powder. The selected powders can then be classified according to their sensibility to tribocharging. We show how the micro properties of the grains as well as the nature of the cohesive interactions drastically influence the charge buildup during the flow. Furthermore, the influence of moisture inside the powder on its electrostatic behaviour is also highlighted. The results give useful insights that will help improving post-manufacturing conditioning of the excipient and/or drug formulation to help to reduce the charge buildup and charge variability.

[1] C. Allenspach, P. Timmins, G. Lumay, J. Holman, T. Minko, International Journal of Pharmaceutics, 596, 2021.

[2] A. Rescaglio, J. Schockmel, F. Francqui, N. Vandewalle, G. Lumay, ANNUAL TRANSACTIONS OF THE NORDIC RHEOLOGY SOCIETY, 25, 2017.

[3] G. Lumay, S. Pillitteri, M. Marck, F. Monsuur, T. Pauly, Q. Ribeyre, F. Francqui, N. Vandewalle, Journal of Drug Delivery Science and Technology, 53, 2019.

627236 Experimental and Numerical Investigations of the RTD in a Gea Consigma CTL25 Tablet Press

Abstract:

The knowledge of the residence time distribution (RTD) of pharmaceutical unit operations is vital for the development of control strategies for continuous manufacturing lines. The present contribution addresses the validation of experimentally determined RTD with numerical results for the tablet press of the GEA ConsiGma CTL25 line.

Various experimental investigations are available which focus on the determination of the RTD in tablet presses via costly tracer experiments (see e.g. [1] and [2]). Due to high computational expenses the numerical examination of the RTD in tablet presses has not been demonstrated before. Previous work [3] showed the potential to calculate the RTD by the means of DEM simulations. We further apply the DEM approach, overcome the computational limitation with a time scale extrapolation method and validate our results with experimental data. Our approach yields the potential to complement experimental DoE runs with numerical simulations and enhance the understanding of material behavior inside this complex unit operation.

Most importantly, the present contribution addresses the following topics: (i) experimental determination of the RTD in a Design-of-Experiment (DoE) wise approach, (ii) characterization and calibration of the investigated material, (iii) the numerical setup similar to prior performed experiments and (iv) validation of the RTD with experimental data.

[1] R. Furukawa, R. Singh, and M. Ierapetritou, 'Effect of material properties on the residence time distribution (RTD) of a tablet press feed frame', Int. J. Pharm., vol. 591, Dec. 2020.

[2] M. Dülle, H. Özcoban, and C. S. Leopold, 'Investigations on the residence time distribution of a three-chamber feed frame with special focus on its geometric and parametric setups', Powder Technol., vol. 331, pp. 276–285, May 2018.

[3] E. Siegmann et al., 'Powder flow and mixing in different tablet press feed frames', Adv. Powder Technol., vol. 31, no. 2, 2020

624321 Agitated Drying of Active Pharmaceutical Ingredients: Effect of Intermittent Mixing on Heat Transfer

Abstract:

Drying is a common unit operation in the upstream processing of active pharmaceutical ingredients (APIs). Although several drying methods have been used in the pharmaceutical industry, agitated drying is a common approach for API drying. During the process, wet particles are heated in a jacketed cylindrical vessel while being mixed by a rotating impeller until the product reaches the desired moisture content. An incomplete understanding of the process can result in operating problems such as under or over drying, non-uniform drying, polymorphism changes, agglomeration, and particle breakage. Moreover, it is difficult to choose the operating conditions to optimize the process due to several phenomena including mass transfer, heat transfer, and changes in material properties which occur simultaneously in an agitated dryer. Even though agitation promotes convective drying, the shear stress acting on the bulk material can result in particle breakage. In previous work, it has been observed that particle breakage is independent of rotation speed and dependent on the number of impeller revolutions.

The discrete element method (DEM) has been used previously to examine heat transfer for particle-particle and particle-wall conduction for a dry bed of particles. The rate of heat transfer can be described by a heating time, which is the time required to heat the bulk material to a fraction of the wall temperature. In this study, we simplified the problem and focused specifically on how to maximize heat transfer while minimizing particle breakage during the drying process. We performed DEM simulations using intermittent mixing, defined as alternate stirring and non-stirring periods of the impeller, as an approach for minimizing particle breakage by decreasing the number of impeller revolutions. We found that the heating time initially decreased with increasing impeller revolutions and then reached a plateau. For a single cycle of intermittent mixing, the starting time of the stirring period influenced the heating time. Moreover, for multiple cycles of intermittent mixing with similar number of impeller revolutions, the heating time of the material bed decreased with increasing intermittent mixing cycles to a certain point and then reached a plateau. When we calculated the relative stirring time (RST), the ratio of stirring time per cycle, we found that intermittent mixing beds with similar RST and number of impeller revolutions gave similar heating time within a certain range of number of impeller revolutions. This work provides insight into how intermittent mixing can be optimized during the drying process in order to maintain effective heat transfer while diminishing particle breakage. This work was partially supported by Takeda Pharmaceuticals International Co.

631004 Beyond Brittle/Ductile Classification: Applying Proper Constitutive Mechanical Metrics to Understand the Compression Characteristics of Pharmaceutical Materials

Abstract:

Pharmaceutical tablet development is complex because of very different objectives, such as a target release profile of the active pharmaceutical ingredient (API), sufficient tablet mechanical strength to limit defects, manufacturability of the powder formulation, and commercial trade dress constraints (e.g., tablet dose strength vs. size requirements) that must be fulfilled. Often, factors that improve one objective can be detrimental to another. For example, a formulation comprising mostly diluents (excipients) may be favored from a manufacturability standpoint. However, it may not be suitable for delivering high doses when a tablet size limit exists. Tablet formulators, then, must be savvy about choosing the right 'ingredients' to optimize all aspects of a tablet.

A key tablet property, its mechanical strength, depends on the compression characteristics of each ingredient that comprises the tablet formulation during and after the tableting process. The tablet characteristic is often classified as 'brittle' or 'ductile' with an elastic/viscoelastic subcategory sometimes added. A ductile tablet is obviously preferred over a brittle one, though a naturally brittle API can still be formulated into a ductile tablet by choosing appropriate diluents. The characteristics of a ductile or brittle material, however, are often misconstrued in the pharmaceutical literature. This is a result of a classification system that's built primarily from correlating the 'in-die' (i.e., during compression) data to out-of-die behaviors of a handful of model materials, e.g., microcrystalline cellulose (MCC) for ductile behavior; mannitol, lactose, and calcium phosphate for brittle behavior; and starch for elastic and viscoelastic behavior. Such a 'one size fits all' approach can subsequently lead to inaccurate classification of APIs, which, more often than not, behave very differently than the above-mentioned model materials.

This study compares the commonly reported mechanical metrics of two proprietary APIs and two classical model excipients: MCC and mannitol. We demonstrate that materials classified as 'ductile' by Heckel's or Kawakita's 'standards' may bond weakly and exhibit the breaking characteristics of a brittle material, meaning that the ductile/brittle behavior under pressure is not always related to the brittle/ductile behavior of the final tablet. In contrast, our data shows that it is possible for materials possessing a high Young's modulus and residual die-wall pressure to break in a ductile manner, akin to compressed metal powders. For the materials studied here, the Hiestand brittle fracture index (BFI) appears to capture the breaking behavior accurately—materials that break in a ductile manner all have small BFIs. Our data further highlight the complexity of pharmaceutical products, in particular APIs, and the need to evaluate a set of mechanical metrics, instead of simplistic assignments of ductility or brittleness based on in-die compression quantities that describe nothing more than the material's compressibility, to classify truly the compression characteristics of tablets.

632057 Computational Modeling of Drying of Pharmaceutical Wet Granules in a Fluidized Bed Dryer Using Coupled CFD-DEM Approach

Abstract:

Introduction/Purpose:

Fluidized bed drying is a topic of interest for scientists and engineers because of its complicated physics and its presence in a wide range of industrial processes e.g. pharmaceutical product manufacturing, food processing, wood processing, reduction of iron ore, flue gas cleaning, the roasting of sulfide ores, drying of coal, catalyst industry and many more. Here, we propose a model based on coupled Computational Fluid Dynamics (CFD) and Discrete Element Method (DEM) approach to simulate the drying of granular particles in fluidized bed dryer associated to the drug manufacturing process. The drying process of wet granular particles in a fluidized bed dryer involves momentum, heat and mass transfer between the particles and the drying medium which is inherently a multiphase-multicomponent flow problem.

Methods:

The model was implemented using an opensource software CFDEM® coupling where the fluid phase was solved by OpenFOAM® codes and the motion of particles were calculated by LIGGGHTS® codes. The information between the fluid phase and particles were exchanged at certain time intervals. A reaction engineering based approach was used to model the evaporation of water from the wet granules and the existing CFDEM® code was modified to incorporate the model for the evaporation. The cohesion force acting between the wet granules due to the formation of liquid bridge between the wet granules was considered in the model.

Results:

Our model was capable of capturing the heat, momentum and mass transfer between the particles and the drying medium. The model was used to predict the drying of lactose monohydrate particles. Numerical simulations were performed to predict the change of particle moisture content and temperature with time during the fluidization process. The variation of average particle temperature and average moisture content with time agreed well with the experimental results obtained from the literature. Agglomeration of wet particles was observed during the drying process due to the cohesion force generated by the liquid bridge between the wet particles. The agglomeration of particles was not present after the particles became dry as expected. It was also observed that the temperature of the dry particles increased at a much higher than the wet particles since the latent heat of evaporation for the water in the wet particles was supplied by the particle itself. This was also observed in the experiments.

Conclusions:

A validated coupled CFD-DEM model was developed to simulate the drying of pharmaceutical wet granules. The effect of particle-particle cohesion due to the formation of liquid bridge between the wet particles was considered in the model. The results obtained from the simulations indicate that the motion and rate of increase of temperature of wet particle are significantly different from those of dry particles. This model can further help set up reliable scale up, troubleshooting or optimization schemes by greatly replacing the burden of the cost of design of experiments.

624081 Influence of Triboelectric Charges on Powder Behaviour in Pharmaceutical Processes

Abstract:

It is well known empirically that electrostatic charges induce problems during powder processing in pharmaceutical processes. Triboelectric charges are produced during powder flow at the contacts between the grains and at the contacts between the grains and the container. The presence of electrostatic charges induces adhesion and cohesiveness leading to intermittent or blocked flow and consequently non-uniformity of dosage. This triboelectric effect is particularly problematic in the framework of continuous manufacturing. For example, material build-up on a loss-in-weight feeder barrel due to powder triboelectric charging results in strong mass flow fluctuations [1]. Moreover, active pharmaceutical ingredients (APIs) are known to be particularly efficient to produce electrostatic charges during flows [2,3].

Unfortunately, the triboelectric effect is still poorly understood, even at the fundamental level. Therefore, the approach to solve practical problems is mostly empirical. Moreover, reproducible electrostatic measurements are difficult to perform. A few years ago, a laboratory instrument (GranuCharge from GranuTools company) has been developed to measure the ability of a powder to produce electrostatic charges during a flow [3]. This instrument allows to predict potential problems induced by electrostatic charges in processes involving powders.

In this presentation, a quick review of the fundamental knowledge regarding the triboelectric effect in powders will be presented. Afterward, the effect on pharmaceutical processes will be discussed. Finally, some solutions to avoid charge build-up will be presented [4].

[1] C. Allenspach, P. Timmins, G. Lumay, J. Holman, and T. Minko, Loss-in-weight feeding, powder flow and electrostatic evaluation for direct compression hydroxypropyl methylcellulose (HPMC) to support continuous manufacturing, International Journal of Pharmaceutics 596, 120259 (2021)

[2] E. Supuk, A. Zarrebini, J. P. Reddy, H. Hughes, M. M. Leane, M. J. Tobyn, P. Timmins, and M. Ghadiri. Tribo-electrification of active pharmaceutical ingredients and excipients, Powder Technology 217, 427 (2012)

[3] A. Rescaglio, F. De Smet, L. Aerts, and G. Lumay, Tribo-electrification of pharmaceutical powder blends, Particulate Science and Technology 37, 1024-1031 (2019)

[4] G. Lumay, S. Pillitteri, M. Marck, F. Monsuur, T. Pauly, Q. Ribeyre, F. Francqui, and N. Vandewalle, Influence of mesoporous silica on powder flow and electrostatic properties on short and long term, Journal of Drug Delivery Science and Technology 53, 101192 (2019)

Triboelectric Charges in Pharmaceutical Processes

Measuring powder tribocharging

GRANUCHARGE™

Effect of tribocharging

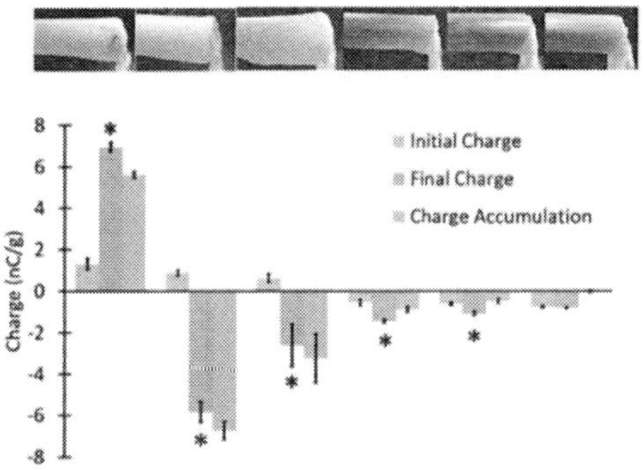

International Journal of Pharmaceutics **596**, 120259 (2021)

Avoiding tribocharging

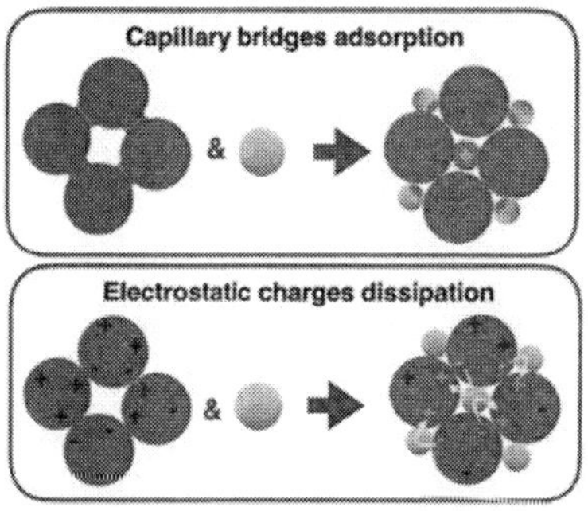

Journal of Drug Delivery Science and Technology **53**, 101192 (2019)

626869 A Detailed Comparison of Comparison of Traditional Pan Coating to Novel Consigma Tablet Coating

Abstract:

Tablet coating is widely used in the pharmaceutical industry. It is typically performed in a batch coating process. In a typical batch coating process, hundreds of kilos of tablets are coated for hours at low rotation rates. The pharmaceutical industry has decades of experience in using batch tablet coater and limited experience for the ConsiGma tablet coater. In recent years continuous manufacturing is increasingly used in the pharmaceutical industry. It promises to deliver high product quality while simultaneously increasing production flexibility. GEA developed the ConsiGma line(s) to be integrated into granular downstream processes. The specially developed tablet coater is unique in its design and process mechanics. It uses high rotation rates compared to traditional drum batch coaters and allows lower process cycle times.

Computational methods are increasingly used in the pharmaceutical industry as well and facilitate a deeper understanding of the process mechanics. The discrete element method (DEM) and computational fluid dynamics (CFD) are two methods that allow the transition from empirical process design to a mechanistic understanding of the individual process units. This work used CFD-DEM simulations to compare a laboratory batch coating process to the coating process inside the ConsiGma tablet coater.

The goal of this work is to present a comprehensive overview of the similarities and differences between the two coating processes. For both coaters, a small design of experiments is performed. The simulations are analyzed for the temperature and coating mass / liquid distribution, forces acting on the tablets during the coating process, cycle time, spray residence time, and spray mass per spray zone visit. The results give a comprehensive overview of the similarities and differences in both coating process types and should simplify the scaling between the two. This work can help accelerate the scale between the two coater types

622354 Accelerated Quantitative DEM Simulation of Large Screw Feeder and Tablet Press Feeder Using Adaptive Coarse Graining Technique

Abstract:

Coarse-graining of particles is unavoidable for practical DEM simulations. For pharmaceutical industry, the typical dimensions of the production scale equipment are ~ 1 m whereas the typical powder contains particles around 0.1 mm. This size ratio requires the coarse-graining ratio to be very high which introduces large inaccuracies in the simulation. To tackle this situation, recently an adaptive coarse-graining strategy has been proposed [1]. This newly proposed technique has been demonstrated using a simple hopper discharge. However, refinement of thc algorithm and application for a variety of systems is pending. In this work we will apply the adaptive coarse-graining technique for two problems of practical importance: tablet die feeder and hopper screw feeder.

It will be shown that the adaptive coarse-grained simulation of tablet die filling reveals additional features that were not otherwise seen in the fixed coarse-grained simulations. This is because the model particles from fixed coarse-graining are much larger than actual particles that introduces artifacts near the die mouth (like exaggerated weight RSD and porosity).Jamming in the feeder section is also observed.As shown in figure 1, resolved particles can be taken near the mouth of the die for accurate estimation of the filling process, when adaptive coarse-graining technique is used. Slightly coarse-grained particles will be taken near the paddle wheel for correct estimation of the flow, and larger coarse-grained particles will be taken to fill up the feed chute. With this strategy, the system can be simulated accurately with reasonable computation.

A different type of challenge exists in implementing adaptive coarsegraining for the hopper screw feeder system. Because of the large recirculation near the feed end of the screw, adaptive coarse-graining induces repeated resolution and consolidation of same set of particles which eats up computational resources and renders adaptive coarse-graining less attractive for such systems. It will be shown that an innovative zoning strategy (snapshot given in figure 1) can resolve this issue and the adaptive coarse-graining can improve the efficiency of such systems substantially.

References:

1. A Multi-Level Coarse Graining Technique for Discrete Element Method (DEM) Simulation. Jayanta Chakraborty, Tarun De, Jitendra Kumar, Salvador García-Muñoz, Maitraye Sen. AIChE Annual Meeting, 2020

Figure 1: Snapshots of the adaptive coarse-grained simulation of tablet die feeder (left) and the screw feeder (right).

625348 Numerical Workflow for Predicting Tablet Coating Uniformity in Pharmaceutical Coating Processes from Lab to Commercial Scale

Abstract:

In this presentation, tablet coating uniformity was predicted in different coater geometries (ProCepT, Glatt GMPC1, Omega and Bohle BFC 5) using commercial software (Rocky DEM and ANSYS CFD). Two tablet shapes () and various operating conditions (drum rotational speed, batch load, spray rate, and nozzle position) were investigated and the results were compared against experimental data. Simulation results showed good agreement with both experimental and analytical models of the coating process, and a sensitivity study was also performed out to determine the influence of the critical parameters during the process. A workflow for predicting the Coefficient of Variation of the tablet film coating thickness using numerical simulation is proposed for virtual process design.

625580 Practical Applications of Dry-Coating for Pharmaceutical Tablet Manufacturing

Abstract:

Dry-coating processes designed to coat API or excipients with glidant (nano-sized silicon dioxide), have been shown effective to improve the manufacturability of formulations containing APIs with poor flowability and tabletability. Despite the advantages of dry-coating at the lab-scale using a wide variety of equipment such as powder blenders/mixers and mills, manufacturing-scale processes have yet to gain significant acceptance. This may be due to the poor scalability of dry-coating processes or presumably due to the aversion of risk associated with adopting new technology in the pharmaceutical industry. To address this issue, this work details the industrial application of dry-coating for use in tablet manufacturing.

In this study, enhancement to powder flowability and tabletability are discussed. Effects and issues related to dissolution as well as sticking/picking during tabletting are also presented. As a major novelty, scale-up studies utilizing a conical screen mill (comill) are detailed. While the comill is frequently employed in the production of tablets as a screening/de-lumping or particle size reduction process, it has not gained acceptance as a dry-coating process. This work shows that the comill can be used as an effective and efficient process for the purpose of dry coating at both lab and manufacturing scale. While our previous work has shown the effectiveness of the comill at the lab scale [1], this work will focus on matters of scale up. The parameters affecting development of the comill as a dry-coating process are examined.

[1] M. Capece, C. Borchardt, A. Jayaraman, Improving the effectiveness of the Comil as a dry-coating process: Enabling direct compaction for high drug loading formulations. Powder Technology 379 (2021) 617-629.

AbbVie funded and participated in the study design, research, data collection, analysis and interpretation of data, as well as writing, reviewing, and approving the publication. Maxx Capece and Jeffery Larson are AbbVie employees and may own AbbVie stock/options.

622638 Scroll Compaction: A Novel Approach for Uniform Compaction of Powder Beds

Abstract:

Consolidation / compaction of a powder bed is a common occurrence in powder handling processes. In many cases, this is intentional, for example during tabletting or isostatic pressing. However, in other scenarios consolidation of a powder is undesirable, leading to poor process performance. Consolidation reduces inter-particular spacing and increases the specific contact area between particles, increasing particle-particle interactions which in turn reduce powder flowability, and in some cases may prevent it altogether, especially when other factors such as environmental conditions are also in effect.

As such, understanding the impact of consolidation is a useful tool for identifying potential processing issues, even where consolidation is not a deliberate process step. Consolidation can occur via many different routes, the most common are vibrational forces, for example during transport, and compaction through the application of applied load, for example during storage where powders may compact under their own mass.

Commonly applied compressibility tests typically use single end compaction to consolidate the powder bed, i.e. the normal stress is applied on a single principal plane. However, consolidation is not homogeneously induced throughout the powder bed using this method, especially with highly compressible materials. One alternative method for preparing a more uniformly compacted powder is to build-up the powder bed layer by layer but this has practical limitations and can be extremely time consuming.

Scroll Compaction is a proposed solution. This approach uses a novel helical blade which rotates upwards through the powder bed, compacting at predefined intervals. Using this method, the operator is only required to fill the test vessel with sufficient material and start the program, no further input is required. This method is also quick; 15-20 compaction steps can be completed in approximately 5 minutes.

To demonstrate the capability of this novel technique, Scroll Compaction was compared with single stage compaction for several materials of varying degrees of cohesion. In order to determine if Scroll Compaction was a more effective method of consolidating materials, the resistance to flow of the compacted bed, presented as a flow energy, was measured alongside the bulk density. Increases in flow energy compared to that of the non-consolidated "as received" sample are indicative of a consolidated powder bed.

The results demonstrate that Scroll Compaction resulted in greater increases in both bulk density and flow energy values, compared with single stage compaction. The effect was more pronounced for cohesive materials, which tend to be less efficient at transmitting forces through the powder bed, however there were still clear differences between the two methodologies for free-flowing, typically non-compressible materials. Overall, the results suggest that Scroll Compaction is more effective for consolidating powders, resulting in a more uniformly, more process-relevant compacted bed.

623555 A DEM-Based Approach for Prediction of Pellet Mixing in a Full-Scale Conical Blender

Abstract:

Blending operations are prevalent in the manufacture of drug products within the pharmaceutical industry, but most often involve powders in diffusion/tumble blenders. In this work, the development of blending process for millimeter-scale pellets in a large-scale, conical screw mixer is discussed. A model-based approach using the discrete element method was used to guide process development. The DEM-based approach was used (1) predict the mixing time required for the full-scale process in a 6000 L conical mixer and (2) guide appropriate selection of blend uniformity acceptance criteria for pellet samples. An experimental mixing study at scale confirmed the mixing conditions and mixing time required to achieve a well-mixed batch.

WK, DM, MD, and BR are employees of AbbVie and may own AbbVie stock. DA is a former employee of AbbVie and may own AbbVie stock. AbbVie sponsored and funded the study; contributed to the design; participated in the collection, analysis, and interpretation of data, and in writing, reviewing, and approval of the final publication.

620597 Using Powder Rheometry to Identify and Quantify Different Humidity Induced Caking Behaviours

Abstract:

"Caking" is the phenomenon where materials gain strength during storage due to changes in particle-particle interactions. Whilst caking occurs via several different mechanisms, for example, mechanical, thermal, environmental and/or chemical, the absorption and migration of moisture is frequently the most dominant mechanism. Fully understanding the propensity to cake is important for minimising downstream process issues, such as blockages, which may have a detrimental impact on the productivity and final product quantity.

It is often assumed that caking occurs uniformly throughout a powder bed, however recent advances in powder rheometry techniques have demonstrated that powders can exhibit a wide range of caking behaviour, including non-homogenous caking (crusting) and even a "hybrid" response dependent on storage conditions. Furthermore, as powder rheometry can quantify the resistance to flow throughout the powder bed, progression of the cake/crust and its resultant strength can be assessed using this method.

To demonstrate the ability of powder rheometry to capture the impact of different caking behaviours, several materials including skimmed milk powder (SMP), sulphated methyl ester (SME) and three food flavours were stored for several days under controlled humidity conditions. SMP exhibited non-homogeneous behaviour with a clearly defined caked region which progressed through the sample over several days. In contrast SME exhibited the aforementioned hybrid behaviour, with the bulk of caking localised, but further caking occurring below the defined crust. In both cases, the strength of the crust was shown to increases over time and demonstrating that caking may require several days to fully develop. Analysis of the food flavours indicated that some materials can also switch between the observed caking regimes, with one material initially exhibiting hybrid behaviour before transitioning to a more homogeneous response.

Caking behaviour may be further complicated by variations in the environmental conditions. To further understand this, samples were stored at 75%RH for 3 days and followed by 3 days at ambient (33%RH). As expected, the return to ambient conditions decreased the degree of caking for some samples. However, for SMP, drying the sample had limited impact on the crusted region but more interestingly, caking continued to occur in the previously uncaked region below the crust. More notably one of the food flavourings presented a significant increase in the degree of caking on return to ambient conditions, demonstrating that for some materials, drying can actually exacerbate caking behaviour.

Overall, the results demonstrated the humidity induced caking is complex, manifesting in a diverse range of different behaviours. In order to fully understand the caking behaviour of a given material, it is therefore necessary to use a technique capable of capturing variations in the flowability throughout the powder bed. Though the use of powder rheometry it is not just possible to identify the caking regime but also quantify the impact, allowing the progression of the cake over an extended period be determined.

629290 Influence of Temperature on the Packing Dynamics of Powders

Abstract:

Numerous industrial processes involve granular materials and fine powders. However, due to the lack of complete theoretical descriptions, the understanding of powder behavior has to rely on the characterization of the macroscopic properties (flow, packing dynamics, agglomeration, segregation, ...). Powder behavior is known to be influenced by particles shape and size distributions, friction forces and cohesive interactions. Cohesive interactions are due to the presence of liquid bridges, electrostatic charges, van der Waals interactions or more rarely magnetic dipole-dipole interactions. For powders, the behavior is usually driven by these cohesive interactions due to the small particle sizes.

Environmental conditions (relative humidity and temperature) can alter the powder properties changing the predominance of these grain interactions. Indeed, moisture is known to influence both surface grains conductivity and capillary bridges formation with as a consequence a modification of the flowing properties. Furthermore, empirical observation seems to indicate that the interactions between the grains as well as grains breakage behavior are affected by temperature. However, precisely controlling the powder temperature during a process requires costly equipment and is almost not feasible for large production facilities. Seasonal changes and production in different sites over the globe induce unavoidable variations of the environmental conditions at which the powder is processed. Therefore, understanding the influence of temperature on the powder behavior is of great interest to predict the degradation of process performance and establish optimal processing conditions. Investigations on the effect of environmental conditions as mainly focused on relative humidity [1-2], temperature has only be considered in processes that involve powder heating at high temperature, such as polymer and metal powders in additive manufacturing [3].

In the present study, we emphasized the importance of even low variations of temperature on common food and lactose powders. Changes in powder properties have been assessed by a packing dynamics characterization method which provides useful information on the powder properties that can be related to its flowability. An improved version of the classical tapped density method has been used (GranuPack, GranuTools)[4] and modified to allow continuous powder heating during the measurement. After each tap, the density is measured accurately with the use of an inductive sensor, allowing to extract useful information on the packing kinematics. Temperature is found to slow down the packing dynamics at temperature variations that are usually considered as normal room fluctuations (20 – 40 °C). The importance of evaluating the effect of temperature even for processes that do not involve powder heating is thus highlighted. We also demonstrate the inability of the classically used Hausner ratio analysis to properly describe this behavior. These new results evidence the need to take into account the environmental conditions at which the powder will be processed at the beginning of the development and characterization stage.

629369 Investigation of Electrostatic Properties of Lactose Powders: Implications for Dry Powder Inhaler (DPI) Processes

Abstract:

Dry powder inhaler (DPI) products are designed to deliver a dry powder active ingredient to a patient's lungs using an inhaler device. The formulation typically consists of a homogenous blend of the active ingredient and an excipient (typically lactose), which is contained within a reservoir, capsule, or blister. Such formulations are produced by physically mixing micronised drug particles and the larger-sized excipient 'carrier' particles. Contrary to many conventional tableting formulations, the excipient is an important functional ingredient in a DPI formulation. The excipient carrier plays at least two roles in the formulation. Since the active ingredient is used in very small quantities, blending with the excipient is necessary to enable delivery of such a small quantity. Second, the excipient must separate from the drug during inhalation so that the drug can be inhaled into the patient's lungs.

The excipient physical properties and interfacial forces affect the blend uniformity and drug release of the DPI formulation. Some of these properties include particle size distribution, flowability, surface chemistry, morphology, and electrostatic properties of the excipient. It is important that these characteristics are controlled during manufacture and storage of the excipient. The main challenges in DPI development are producing a uniform blend for accurate dosing and then proper drug separation during inhalation. Beyond that, reducing batch to batch variability is extremely challenging in these types of formulations.

Electrostatic charges of the excipient and drug particles (API) can affect the performance of dry powder inhalation drug formulations. Electrostatic forces on the excipient could cause the API particle to adhere to the surface of the excipient and prevent release. Also, electrostatic forces in the final blend could affect the powders release from the device. Of course, variability in charge profiles could cause variability in the final drug fine particle fraction (API that would be released to the patient). Characterization of these charges is an important tool to possibly allow a higher percentage of the API to be released from the excipient/device. Managing or controlling these charges could also be important to reduce the variability of API release. However, due to the complexity of the mechanisms at the origin of triboelectric effects, electrostatics has been only recently investigated for pharmaceutical powders [1,2].

In this study, the electrostatic properties of lactose excipients designed for DPI applications (AeroFlo, Kerry) have been investigated with the GranuCharge (GranuTools)[3]. Anhydrous and monohydrate powders differing from their particle size distribution have been selected. Moisture is controlled by conditioning the powders before the measurements at a fixed relative humidity with a specifically designed rotating tumbler coupled with a humid air generator (GranuMidity, Granutools). Then, powder charge density is measured before and after the powder flow through a Stainless Steel SS316L pipe. During the flow, the particle/particle and particle/pipe interactions lead to a charge buildup due to triboelectric effect, increasing the overall charge density of the powder. The selected powders can then be classified according to their sensibility to tribocharging. We show how the micro properties of the grains as well as the nature of the cohesive interactions drastically influence the charge buildup during the flow. Furthermore, the influence of moisture inside the powder on its electrostatic behaviour is also highlighted. The results give useful insights that will help improving post-manufacturing conditioning of the excipient and/or drug formulation to help to reduce the charge buildup and charge variability.

[1] C. Allenspach, P. Timmins, G. Lumay, J. Holman, T. Minko, International Journal of Pharmaceutics, 596, 2021.

[2] A. Rescaglio, J. Schockmel, F. Francqui, N. Vandewalle, G. Lumay, ANNUAL TRANSACTIONS OF THE NORDIC RHEOLOGY SOCIETY, 25, 2017.

[3] G. Lumay, S. Pillitteri, M. Marck, F. Monsuur, T. Pauly, Q. Ribeyre, F. Francqui, N. Vandewalle, Journal of Drug Delivery Science and Technology, 53, 2019.

627236 Experimental and Numerical Investigations of the RTD in a Gea Consigma CTL25 Tablet Press

Abstract:

The knowledge of the residence time distribution (RTD) of pharmaceutical unit operations is vital for the development of control strategies for continuous manufacturing lines. The present contribution addresses the validation of experimentally determined RTD with numerical results for the tablet press of the GEA ConsiGma CTL25 line.

Various experimental investigations are available which focus on the determination of the RTD in tablet presses via costly tracer experiments (see e.g. [1] and [2]). Due to high computational expenses the numerical examination of the RTD in tablet presses has not been demonstrated before. Previous work [3] showed the potential to calculate the RTD by the means of DEM simulations. We further apply the DEM approach, overcome the computational limitation with a time scale extrapolation method and validate our results with experimental data. Our approach yields the potential to complement experimental DoE runs with numerical simulations and enhance the understanding of material behavior inside this complex unit operation.

Most importantly, the present contribution addresses the following topics: (i) experimental determination of the RTD in a Design-of-Experiment (DoE) wise approach, (ii) characterization and calibration of the investigated material, (iii) the numerical setup similar to prior performed experiments and (iv) validation of the RTD with experimental data.

[1] R. Furukawa, R. Singh, and M. Ierapetritou, 'Effect of material properties on the residence time distribution (RTD) of a tablet press feed frame', Int. J. Pharm., vol. 591, Dec. 2020.

[2] M. Dülle, H. Özcoban, and C. S. Leopold, 'Investigations on the residence time distribution of a three-chamber feed frame with special focus on its geometric and parametric setups', Powder Technol., vol. 331, pp. 276–285, May 2018.

[3] E. Siegmann et al., 'Powder flow and mixing in different tablet press feed frames', Adv. Powder Technol., vol. 31, no. 2, 2020

624321 Agitated Drying of Active Pharmaceutical Ingredients: Effect of Intermittent Mixing on Heat Transfer

Abstract:

Drying is a common unit operation in the upstream processing of active pharmaceutical ingredients (APIs). Although several drying methods have been used in the pharmaceutical industry, agitated drying is a common approach for API drying. During the process, wet particles are heated in a jacketed cylindrical vessel while being mixed by a rotating impeller until the product reaches the desired moisture content. An incomplete understanding of the process can result in operating problems such as under or over drying, non-uniform drying, polymorphism changes, agglomeration, and particle breakage. Moreover, it is difficult to choose the operating conditions to optimize the process due to several phenomena including mass transfer, heat transfer, and changes in material properties which occur simultaneously in an agitated dryer. Even though agitation promotes convective drying, the shear stress acting on the bulk material can result in particle breakage. In previous work, it has been observed that particle breakage is independent of rotation speed and dependent on the number of impeller revolutions.

The discrete element method (DEM) has been used previously to examine heat transfer for particle-particle and particle-wall conduction for a dry bed of particles. The rate of heat transfer can be described by a heating time, which is the time required to heat the bulk material to a fraction of the wall temperature. In this study, we simplified the problem and focused specifically on how to maximize heat transfer while minimizing particle breakage during the drying process. We performed DEM simulations using intermittent mixing, defined as alternate stirring and non-stirring periods of the impeller, as an approach for minimizing particle breakage by decreasing the number of impeller revolutions. We found that the heating time initially decreased with increasing impeller revolutions and then reached a plateau. For a single cycle of intermittent mixing, the starting time of the stirring period influenced the heating time. Moreover, for multiple cycles of intermittent mixing with similar number of impeller revolutions, the heating time of the material bed decreased with increasing intermittent mixing cycles to a certain point and then reached a plateau. When we calculated the relative stirring time (RST), the ratio of stirring time per cycle, we found that intermittent mixing beds with similar RST and number of impeller revolutions gave similar heating time within a certain range of number of impeller revolutions. This work provides insight into how intermittent mixing can be optimized during the drying process in order to maintain effective heat transfer while diminishing particle breakage. This work was partially supported by Takeda Pharmaceuticals International Co.

631004 Beyond Brittle/Ductile Classification: Applying Proper Constitutive Mechanical Metrics to Understand the Compression Characteristics of Pharmaceutical Materials

Abstract:

Pharmaceutical tablet development is complex because of very different objectives, such as a target release profile of the active pharmaceutical ingredient (API), sufficient tablet mechanical strength to limit defects, manufacturability of the powder formulation, and commercial trade dress constraints (e.g., tablet dose strength vs. size requirements) that must be fulfilled. Often, factors that improve one objective can be detrimental to another. For example, a formulation comprising mostly diluents (excipients) may be favored from a manufacturability standpoint. However, it may not be suitable for delivering high doses when a tablet size limit exists. Tablet formulators, then, must be savvy about choosing the right 'ingredients' to optimize all aspects of a tablet.

A key tablet property, its mechanical strength, depends on the compression characteristics of each ingredient that comprises the tablet formulation during and after the tableting process. The tablet characteristic is often classified as 'brittle' or 'ductile' with an elastic/viscoelastic subcategory sometimes added. A ductile tablet is obviously preferred over a brittle one, though a naturally brittle API can still be formulated into a ductile tablet by choosing appropriate diluents. The characteristics of a ductile or brittle material, however, are often misconstrued in the pharmaceutical literature. This is a result of a classification system that's built primarily from correlating the 'in-die' (i.e., during compression) data to out-of-die behaviors of a handful of model materials, e.g., microcrystalline cellulose (MCC) for ductile behavior; mannitol, lactose, and calcium phosphate for brittle behavior; and starch for elastic and viscoelastic behavior. Such a 'one size fits all' approach can subsequently lead to inaccurate classification of APIs, which, more often than not, behave very differently than the above-mentioned model materials.

This study compares the commonly reported mechanical metrics of two proprietary APIs and two classical model excipients: MCC and mannitol. We demonstrate that materials classified as 'ductile' by Heckel's or Kawakita's 'standards' may bond weakly and exhibit the breaking characteristics of a brittle material, meaning that the ductile/brittle behavior under pressure is not always related to the brittle/ductile behavior of the final tablet. In contrast, our data shows that it is possible for materials possessing a high Young's modulus and residual die-wall pressure to break in a ductile manner, akin to compressed metal powders. For the materials studied here, the Hiestand brittle fracture index (BFI) appears to capture the breaking behavior accurately—materials that break in a ductile manner all have small BFIs. Our data further highlight the complexity of pharmaceutical products, in particular APIs, and the need to evaluate a set of mechanical metrics, instead of simplistic assignments of ductility or brittleness based on in-die compression quantities that describe nothing more than the material's compressibility, to classify truly the compression characteristics of tablets.

632057 Computational Modeling of Drying of Pharmaceutical Wet Granules in a Fluidized Bed Dryer Using Coupled CFD-DEM Approach

Abstract:

Introduction/Purpose:

Fluidized bed drying is a topic of interest for scientists and engineers because of its complicated physics and its presence in a wide range of industrial processes e.g. pharmaceutical product manufacturing, food processing, wood processing, reduction of iron ore, flue gas cleaning, the roasting of sulfide ores, drying of coal, catalyst industry and many more. Here, we propose a model based on coupled Computational Fluid Dynamics (CFD) and Discrete Element Method (DEM) approach to simulate the drying of granular particles in fluidized bed dryer associated to the drug manufacturing process. The drying process of wet granular particles in a fluidized bed dryer involves momentum, heat and mass transfer between the particles and the drying medium which is inherently a multiphase-multicomponent flow problem.

Methods:

The model was implemented using an opensource software CFDEM® coupling where the fluid phase was solved by OpenFOAM® codes and the motion of particles were calculated by LIGGGHTS® codes. The information between the fluid phase and particles were exchanged at certain time intervals. A reaction engineering based approach was used to model the evaporation of water from the wet granules and the existing CFDEM® code was modified to incorporate the model for the evaporation. The cohesion force acting between the wet granules due to the formation of liquid bridge between the wet granules was considered in the model.

Results:

Our model was capable of capturing the heat, momentum and mass transfer between the particles and the drying medium. The model was used to predict the drying of lactose monohydrate particles. Numerical simulations were performed to predict the change of particle moisture content and temperature with time during the fluidization process. The variation of average particle temperature and average moisture content with time agreed well with the experimental results obtained from the literature. Agglomeration of wet particles was observed during the drying process due to the cohesion force generated by the liquid bridge between the wet particles. The agglomeration of particles was not present after the particles became dry as expected. It was also observed that the temperature of the dry particles increased at a much higher than the wet particles since the latent heat of evaporation for the water in the wet particles was supplied by the particle itself. This was also observed in the experiments.

Conclusions:

A validated coupled CFD-DEM model was developed to simulate the drying of pharmaceutical wet granules. The effect of particle-particle cohesion due to the formation of liquid bridge between the wet particles was considered in the model. The results obtained from the simulations indicate that the motion and rate of increase of temperature of wet particle are significantly different from those of dry particles. This model can further help set up reliable scale up, troubleshooting or optimization schemes by greatly replacing the burden of the cost of design of experiments.

624081 Influence of Triboelectric Charges on Powder Behaviour in Pharmaceutical Processes

Abstract:

It is well known empirically that electrostatic charges induce problems during powder processing in pharmaceutical processes. Triboelectric charges are produced during powder flow at the contacts between the grains and at the contacts between the grains and the container. The presence of electrostatic charges induces adhesion and cohesiveness leading to intermittent or blocked flow and consequently non-uniformity of dosage. This triboelectric effect is particularly problematic in the framework of continuous manufacturing. For example, material build-up on a loss-in-weight feeder barrel due to powder triboelectric charging results in strong mass flow fluctuations [1]. Moreover, active pharmaceutical ingredients (APIs) are known to be particularly efficient to produce electrostatic charges during flows [2,3].

Unfortunately, the triboelectric effect is still poorly understood, even at the fundamental level. Therefore, the approach to solve practical problems is mostly empirical. Moreover, reproducible electrostatic measurements are difficult to perform. A few years ago, a laboratory instrument (GranuCharge from GranuTools company) has been developed to measure the ability of a powder to produce electrostatic charges during a flow [3]. This instrument allows to predict potential problems induced by electrostatic charges in processes involving powders.

In this presentation, a quick review of the fundamental knowledge regarding the triboelectric effect in powders will be presented. Afterward, the effect on pharmaceutical processes will be discussed. Finally, some solutions to avoid charge build-up will be presented [4].

[1] C. Allenspach, P. Timmins, G. Lumay, J. Holman, and T. Minko, Loss-in-weight feeding, powder flow and electrostatic evaluation for direct compression hydroxypropyl methylcellulose (HPMC) to support continuous manufacturing, International Journal of Pharmaceutics 596, 120259 (2021)

[2] E. Supuk, A. Zarrebini, J. P. Reddy, H. Hughes, M. M. Leane, M. J. Tobyn, P. Timmins, and M. Ghadiri. Tribo-electrification of active pharmaceutical ingredients and excipients, Powder Technology 217, 427 (2012)

[3] A. Rescaglio, F. De Smet, L. Aerts, and G. Lumay, Tribo-electrification of pharmaceutical powder blends, Particulate Science and Technology 37, 1024-1031 (2019)

[4] G. Lumay, S. Pillitteri, M. Marck, F. Monsuur, T. Pauly, Q. Ribeyre, F. Francqui, and N. Vandewalle, Influence of mesoporous silica on powder flow and electrostatic properties on short and long term, Journal of Drug Delivery Science and Technology 53, 101192 (2019)

Triboelectric Charges in Pharmaceutical Processes

Measuring powder tribocharging

GRANUCHARGE™

Effect of tribocharging

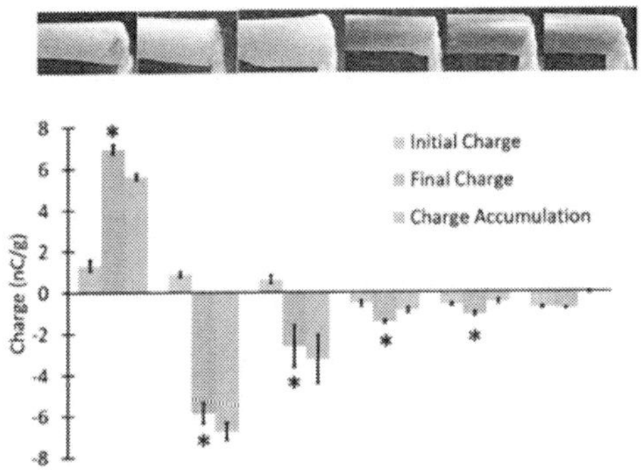

International Journal of Pharmaceutics **596**, 120259 (2021)

Avoiding tribocharging

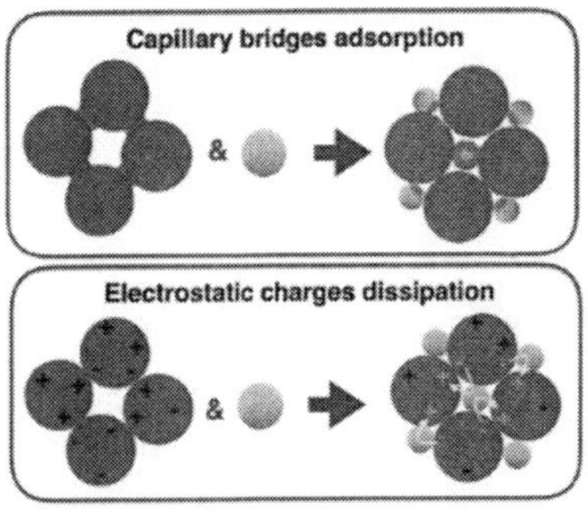

Journal of Drug Delivery Science and Technology **53**, 101192 (2019)

626869 A Detailed Comparison of Comparison of Traditional Pan Coating to Novel Consigma Tablet Coating

Abstract:

Tablet coating is widely used in the pharmaceutical industry. It is typically performed in a batch coating process. In a typical batch coating process, hundreds of kilos of tablets are coated for hours at low rotation rates. The pharmaceutical industry has decades of experience in using batch tablet coater and limited experience for the ConsiGma tablet coater. In recent years continuous manufacturing is increasingly used in the pharmaceutical industry. It promises to deliver high product quality while simultaneously increasing production flexibility. GEA developed the ConsiGma line(s) to be integrated into granular downstream processes. The specially developed tablet coater is unique in its design and process mechanics. It uses high rotation rates compared to traditional drum batch coaters and allows lower process cycle times.

Computational methods are increasingly used in the pharmaceutical industry as well and facilitate a deeper understanding of the process mechanics. The discrete element method (DEM) and computational fluid dynamics (CFD) are two methods that allow the transition from empirical process design to a mechanistic understanding of the individual process units. This work used CFD-DEM simulations to compare a laboratory batch coating process to the coating process inside the ConsiGma tablet coater.

The goal of this work is to present a comprehensive overview of the similarities and differences between the two coating processes. For both coaters, a small design of experiments is performed. The simulations are analyzed for the temperature and coating mass / liquid distribution, forces acting on the tablets during the coating process, cycle time, spray residence time, and spray mass per spray zone visit. The results give a comprehensive overview of the similarities and differences in both coating process types and should simplify the scaling between the two. This work can help accelerate the scale between the two coater types

621312 Experimental Investigation of Route-Mixing for Non-Spherical Particles in the Pneumatic Conveying System

Abstract:

This study aims to investigate the route-mixing inside the horizontal column through the pneumatic conveying system. Different types of non-spherical particles, such as ellipsoids (prolate and oblate), cube, and elongated needles, have been used. Several experiments are performed to examine the effects of parameters such as size, shape, and density on segregation. The flow behavior and orientations of particles have been captured through a high-speed camera. The segregation pattern is observed as decreasing order for the coarse particle as Oblate > Prolate > Elongated needle. The drag force decreases as an increased aspect ratio of non-spherical particles, as shown in Figure 1. Outcomes indicate that on increasing the angularity of coarse particles, the overall segregation index is reduced. The pneumatic conveying system can be used as an alternate option for mixing in the process industry.

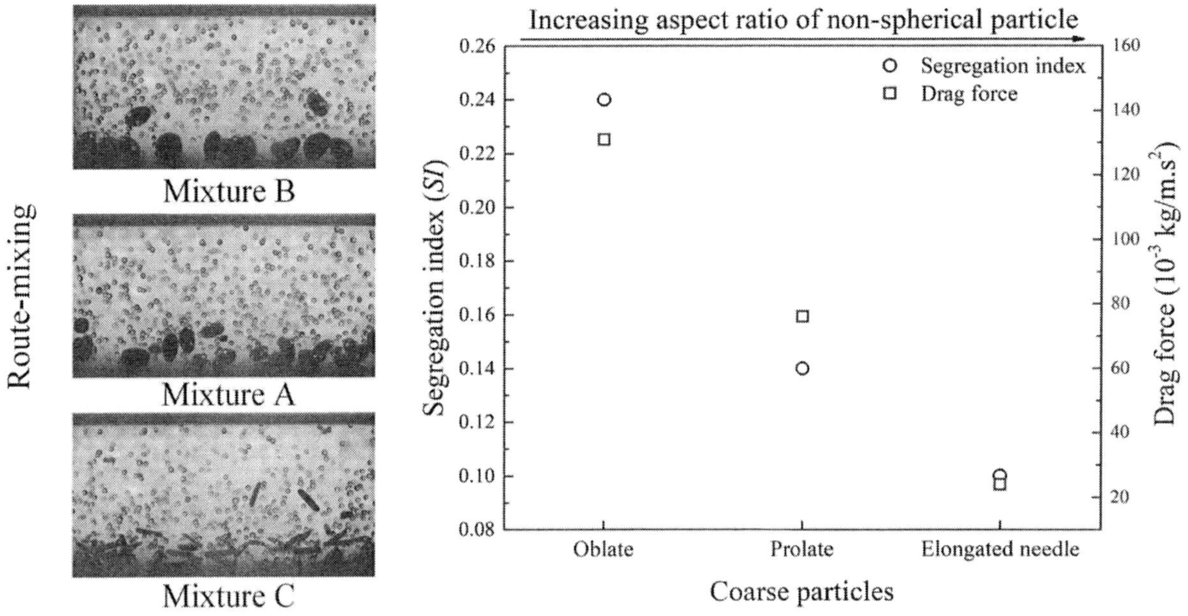

Figure 1: Comparison of segregation index for different coarse particle shape

627393 A Comprehensive Model for the CFD Simulation of Autothermal Fast Pyrolysis of Biomass

Abstract:

The thermochemical conversion of biomass to biofuels via fast pyrolysis has been of growing interest in recent decades. The conventional method is done in the absence of oxygen, and the pyrolysis of biomass is an endothermic process. The heat needed for the process is then supplied from the walls of the reactor. While supplying heat in this manner may be possible for small-scale reactors, it quickly becomes unfeasible for plant-scale reactors due to the reduced surface area to volume ratio of the reactor. Heat transfer becomes a bottleneck for the scale-up of the conventional fast pyrolysis of biomass[1,2].

Autothermal biomass pyrolysis processes[1] are being developed to address the bottleneck to biomass pyrolysis scale-up. Small amounts of oxygen are injected into the fluidized bed reactor, which allows for partial oxidation[1,3] of pyrolysis products. The exothermic reactions supply heat for the endothermic reactions, such as devolatilization, balancing the energy requirements for the process. One of the major contributors of heat in the fluidized bed is char combustion, and the retention of char in the bed is essential for its reaction with oxygen. Simultaneously, the quick release of volatile gases from the biomass is essential for bio-oil yield. The performance of autothermal pyrolyzers may be affected by several design choices of the fluidized bed reactor. For example, the height of the injection point of biomass may influence the mixing of biomass and sand in the reactor, affecting the residence time of the biomass and consequently the product yield.

To investigate these effects, we have developed a computational model to describe biomass fast pyrolysis. The model is based on the Eulerian multi-fluid model[4,5], which originates from a random field statistical approach for gas-solid glows. The phases are treated as interpenetrating continua that obey conservations of mass, momentum, and energy. The biomass and sand are granular phases, and the closure for solid granular properties such as pressure, viscosity, and energy are obtained from the kinetic theory of polydisperse granular flows[6,7] and frictional stress models[8,9]. The model is implemented into OpenFOAM®. A comprehensive chemical kinetics mechanism[10–13] to account for devolatilization, char combustion, and secondary gas-phase reactions was incorporated into the solver.

Initially, the hydrodynamics and the chemical kinetic mechanism were studied independently. The multiphase solver was used to study the mixing of biomass and sand in the fluidized bed. Three-dimensional cold-flow simulations of a lab-scale fluidized bed reactor without reactions were done. Two designs with different biomass injection locations are considered to study the impact on mixing of the biomass injection height. A new type of mixing index based on the average volume fraction field was developed to quantify the mixing from the multiphase CFD simulations. The mixing indices developed are fields that provide the spatial representation of gas-solid mixing and solid-solid mixing in the fluidized bed. These new indices prove to be a powerful tool to identify regions of rich and lean mixing. Conventional mixing indices such as Lacey and Rose are extended to be used with the new mixing index and provide a single scalar measure of mixing for the entire fluidized bed.

Two chemical kinetic mechanisms were taken into consideration, namely, the Miller-Bellan[10] and the Ranzi[11,12] mechanisms. The Ranzi model is realistic because it has the information of gas species produced from the devolatilization of biomass. It consists of two aspects, (a) the devolatilization reactions, which account for the volatile gases and char that are first obtained from biomass, and (b) the secondary gas-phase reactions, which comprise the CRECK model where the volatile gases further react. The CRECK secondary gas-phase model has 137 species and 4533 reactions. The Miller-Bellan mechanism is a simplified mechanism, where gas species are lumped into single species for non-condensable gases and bio-oil vapors. Both mechanisms were developed to study multi-component biomass feedstocks made up of cellulose, hemicellulose, and lignin. Both mechanisms follow a similar pattern where first, the virgin biomass forms an activated component, often considered a depolymerization step. Then volatile gases and char are produced from these activated biomass species. Char combustion reactions were added to these mechanisms to account for the main heat source in the autothermal pyrolysis process.

The chemical kinetic models were implemented in the chemistry solver available in OpenFOAM. The time evolution of the species and heat of reactions were studied. The Miller-Bellan mechanism was modified to match the performance of the Ranzi mechanism. Isothermal, zero-dimensional simulations of the chemistry were performed for various temperatures to study both mechanisms. Different biomass feedstocks were considered, but the focus was on red oak and corn stover.

The chemical kinetic models were incorporated into the multiphase solver with the fluidized bed reactor. The lab-scale reactor has a diameter of 3.81 cm and is 42.7 cm in height. Sand was filled to a height of 10.5 cm, and the bed is initially stationary at a fixed initial temperature. Biomass is injected into the bed at 300 K from the reactor side, while the fluidizing gas enters from underneath the bed. The fluidizing gas comprises a mixture of air and nitrogen injected with a superficial gas velocity at the inlet. Biomass is injected at a rate of 1 kg/h, while the fluidizing gas enters the domain at 20 SLPM. Unlike the chemistry solver, the CFD simulations were not isothermal, but varying initial thermal conditions of the bed are considered as sensible heat is needed to raise the temperature of biomass form 300K to the desired pyrolysis temperature. The yield of bio-oil is computed for these conditions and contrasted with the conventional fast pyrolysis process.

The Ranzi mechanism, being an elaborate chemical kinetic model coupled with the multiphase CFD solver, is computationally expensive. The feasibility of such a comprehensive model as a predictive tool for the yields of bio-oils in the pyrolysis process comes under scrutiny. A plug flow reactor (PFR) model of the lab-scale pyrolyzer was used and showed little to no secondary gas-phase reactions with the CRECK mechanism for the size and operating conditions of the reactor. However, the secondary gas-phase reaction rates are low and depend on the residence time of the gas species in the fluidized bed reactor, which in turn depends on the solid-gas mixing. Therefore, based on the Damköhler number, which is the ratio of mixing to chemical reaction time scales, the secondary gas-phase reactions may be present in the reactor and captured by the CFD model. If these reactions are prevalent in the reactor, they should not be excluded from the model, as the heat from the reactions of the gas species will contribute to the autothermal pyrolysis process.

References

1. Polin JP, Peterson CA, Whitmer LE, Smith RG, Brown RC. Process intensification of biomass fast pyrolysis through autothermal operation of a fluidized bed reactor. *Appl Energy.* 2019;249:276-285. doi:10.1016/j.apenergy.2019.04.154
2. Proano-Aviles J, Lindstrom JK, Johnston PA, Brown RC. Heat and Mass Transfer Effects in a Furnace-Based Micropyrolyzer. *Energy Technol.* 2017;5(1):189-195. doi:10.1002/ente.201600279

3. Kim KH, Bai X, Rover M, Brown RC. The effect of low-concentration oxygen in sweep gas during pyrolysis of red oak using a fluidized bed reactor. *Fuel*. 2014;124:49-56. doi:10.1016/j.fuel.2014.01.086
4. Drew DA. Averaged equations for two-phase flows. *Stud Appl Math*. 1971;L(3):205-231.
5. Drew DA. Continuum Modeling of Two-Phase Flows. In: Meyer R, ed. *Theory of Dispersed Multiphase Flow*. Academic Press; 1983:173-190.
6. Jenkins JT, Savage SB. A theory of the rapid flow of identical, smooth, nearly elastic spherical particles. *J Fluid Mech*. 1983;(130):187-202.
7. Jenkins JT, Mancini F. Balance Laws and Constitutive Relations for Plane Flows of a Dense, Binary Mixture of Smooth, Nearly Elastic, Circular Disks. *J Appl Mech*. 1987;54(1):27-34. doi:10.1115/1.3172990
8. Johnson PC, Jackson R. Frictional–collisional constitutive relations for granular materials, with application to plane shearing. *J Fluid Mech*. 1987;176:67-93. doi:10.1017/S0022112087000570
9. Srivastava A, Sundaresan S. Analysis of a frictional–kinetic model for gas–particle flow. *Powder Technol*. 2003;129(1):72-85. doi:10.1016/S0032-5910(02)00132-8
10. Miller RS, Bellan J. A generalized biomass pyrolysis model based on superimposed cellulose, hemicellulose and lignin kinetics. *Combust Sci Technol*. 1997;126(1-6):97-137. doi:10.1080/00102209708935670
11. Ranzi E, Debiagi PEA, Frassoldati A. Mathematical Modeling of Fast Biomass Pyrolysis and Bio-Oil Formation. Note I: Kinetic Mechanism of Biomass Pyrolysis. *ACS Sustain Chem Eng*. 2017;5(4):2867-2881. doi:10.1021/acssuschemeng.6b03096
12. Ranzi E, Debiagi PEA, Frassoldati A. Mathematical Modeling of Fast Biomass Pyrolysis and Bio-Oil Formation. Note II: Secondary Gas-Phase Reactions and Bio-Oil Formation. *ACS Sustain Chem Eng*. 2017;5(4):2882-2896. doi:10.1021/acssuschemeng.6b03098
13. Calonaci M, Grana R, Barker Hemings E, Bozzano G, Dente M, Ranzi E. Comprehensive Kinetic Modeling Study of Bio-oil Formation from Fast Pyrolysis of Biomass. *Energy Fuels*. 2010;24(10):5727-5734. doi:10.1021/ef1008902

622760 An Index to Characterize Mixing from Average Volume Fraction Fields in Gas Solid Flows

Abstract:

Particle mixing and segregation are common processes used in many chemical engineering applications. For example, in chemical looping combustion[1], gas-solid mixing plays an important role when gas-solid reactions are present. In the fast pyrolysis of biomass[2], solid mixing in fluidized bed reactors is crucial where biomass mixes with sand which acts as the heat carrier. Characterizing mixing and segregation of particles that vary in size and density have been extensively studied for many decades.

Non-invasive experimental studies such as X-ray tomography-computed tomography[3,4] have been conducted for flow visualization within the interior of granular mixers. Granular mixing simulations can be done by using Discrete Element Methods (DEM)[5,6], where the motion of each particle is tracked, or granular flow can be modeled using a continuum approach that obeys Eulerian conservation equations of mass, momentum, and energy. The multiphase computational fluid dynamic (mCFD) models are multi-fluid models based on a random field statistical approach[7,8] to gas-solid flows. For systems with many particles, DEM simulations become infeasible as the computational expense increases with particles, but the Eulerian model is attractive.

A mixing index is often used to quantify solid mixing, and there are many mixing indices proposed over several decades[9–11]. The most widely used index is Lacey's[12] index which gives the ratio of how much mixing has occurred to how much could occur. It is computed by measuring the departure of the sample variance in the fraction of particles of interest at a given state from the completely segregated system normalized by the difference in the fraction between completely segregated and perfectly mixed systems. There are several variants of the Lacey index, such as Kramer[13] and Rose[9], that use the standard deviation instead of the variance.

Most mixing indices are based on particle data obtained from either experiments or DEM simulations, and these methods correspond to one realization of particle positions. However, multiphase computational fluid dynamics of gas-solid flow are based on averaged field representations. Therefore, the characterization of mixing in a particle realization is different from characterizing it in an averaged field representation. The question of the applicability of Lacey's index to mCFD is then not straightforward. Mixing in multiphase CFD is based on average volume fraction fields, and an index must be based on continuous fields defined on a grid.

Two new mixing indices are developed and introduced for gas-solid mixing based on average solid volume fraction fields. These mixing indices are the richness index that characterizes gas-solid mixing, and the solid fraction mixing index that characterizes solid-solid mixing. The index for gas-solid mixing is based on the particle volume fraction fields, and the index for solid-solid mixing is based on solid fraction fields. In a polydisperse system of particles, the *solid fraction* of a particle species is defined as its volume fraction divided by the total particle volume fraction. The idea of the new index is analogous to the mixture fraction from single-phase mixing of fuel and oxidizer in combustion.

The richness index normalizes gas-solid mixing characteristics spatially by performing a local linear transformation and by eliminating dependence on the global average of the total solid volume fraction. Therefore, the richness mixing index scales the volume fraction field between 0 (pure gas) and 1 (fully packed solids), indicating the local departure from perfect mixing corresponding to 0.5. The solid fraction mixing index, , employs a similar idea. It scales the solid fraction field between 0 (absence of particle A) and 1 (only particle A is present), indicating the local departure from perfect mixing corresponding to 0.5, thereby eliminating the dependence on the global average of the solid fraction of individual solid particle types. Perfect gas-solid mixing corresponds to the local total solid volume fraction being equal to the global average total solid volume fraction. In contrast, perfect solid-solid mixing corresponds to the local solid fraction of solid type A being equal to the global average solid fraction of A. These mixing indices are fields that capture spatial mixing patterns in a device. They are, in that sense, local and are hence called *local mixing indices*.

The performance of the new mixing indices is tested on two fluidized bed applications. The first application is a three-dimensional simulation of the mixing of biomass and sand in a fluidized bed reactor designed for fast pyrolysis. Two designs with different biomass injection locations are evaluated to determine whether a higher or lower injection location for the biomass yields better mixing. The second application is a two-dimensional simulation of segregation of an initially perfectly mixed binary mixture of particles with similar densities but different sizes. Simulations of the two applications are performed with the same solver in the OpenFOAM® computational toolbox for fluid dynamics.

The mixing indices are computed from a grid of sampling volumes. The choice of the size of the sampling volume depends on the relevant length scales of the system. There are three characteristic length scales: (a) The physical length scale, which corresponds to the solid volume fraction field, (b) the computational length scale that corresponds to the mesh size on which the CFD simulations are performed, and (c) the length scale corresponding to the sampling volume. The mixing indices can be applied directly to the same computational mesh used for CFD, which implies that the sampling volume length scale is equal to the computational length scale. However, larger sampling volumes could be used. For sampling volumes larger than the computational length scale but smaller than the smallest physical length scale, we expect the local mixing indices to be independent of sampling volume size.

Mixing can vary depending on the scale at which the system is observed and can be characterized by a sampling volume. If the sampling volumes larger than the smallest physical length scales are chosen, micro-mixing features may be excluded, and mixing at larger scales can be studied. The new mixing indices account for the phenomena of scale-dependent mixing. The performance of scale-dependence of the mixing index is characterized using synthetic fields as the data obtained from multi-fluid CFD simulations.

The conventional mixing indices such as Lacey and Rose are also computed from the sampling volume grid. They provide a single scalar measure of mixing in a system and can be called *global mixing indices*. We demonstrate the use of these indices with solid volume fraction and the solid fractions for gas-solid and solid-solid mixing, respectively. We extend these variance-based indices to be used with the local mixing index fields and find considerable improvement when using the global Rose solid mixing index, as opposed to the Rose solid fraction index.

A new performance criterion called the sensitivity of a mixing index is introduced. The local mixing indices developed have tunable sensitivity. Depending on the application, only a portion of the range of solid volume fraction or solid fraction can be accessed. By choosing an adequate accessed region, the mixing index can capture more details of local relative mixing and segregation. Consequently, the sensitivity of global indices based on the standard deviation or variance of the local solid mixing index is influenced. The sensitivity of the global indices can also be influenced by the choice of reference distributions corresponding to the perfectly mixed and perfectly unmixed states.

References

1. Mahalatkar K, Kuhlman J, Huckaby ED, O'Brien T. CFD simulation of a chemical-looping fuel reactor utilizing solid fuel. *Chem Eng Sci.* 2011;66(16):3617-3627. doi:10.1016/j.ces.2011.04.025
2. Bridgwater AV, Meier D, Radlein D. An overview of fast pyrolysis of biomass. *Org Geochem.* 1999;30(12):1479-1493. doi:10.1016/S0146-6380(99)00120-5
3. Franka NP, Heindel TJ. Local time-averaged gas holdup in a fluidized bed with side air injection using X-ray computed tomography. *Powder Technol.* 2009;193(1):69-78. doi:10.1016/j.powtec.2009.02.008
4. Mudde RF. Time-resolved X-ray tomography of a fluidized bed. *Powder Technol.* 2010;199(1):55-59. doi:10.1016/j.powtec.2009.04.021
5. Chandratilleke GR, Yu AB, Bridgwater J. A DEM study of the mixing of particles induced by a flat blade. *Chem Eng Sci.* 2012;79:54-74. doi:10.1016/j.ces.2012.05.010
6. Bhalode P, Ierapetritou M. A review of existing mixing indices in solid-based continuous blending operations. *Powder Technol.* 2020;373:195-209. doi:10.1016/j.powtec.2020.06.043
7. Drew DA. Mathematical Modeling of Two-Phase Flow. *Annu Rev Fluid Mech.* 1983;15(1):261-291. doi:10.1146/annurev.fl.15.010183.001401
8. Pai MG, Subramaniam S. A comprehensive probability density function formalism for multiphase flows. *J Fluid Mech.* 2009;628:181-228. doi:10.1017/S002211200900617X
9. Williams JC. The mixing of dry powders. *Powder Technol.* 1968;2(1):13-20. doi:10.1016/0032-5910(68)80028-2
10. Fan LT, Chen SJ, Watson CA. ANNUAL REVIEW Solids Mixing. doi:10.1021/ie50727a009
11. Bridgwater J. Mixing of powders and granular materials by mechanical means—A perspective. *Particuology.* 2012;10(4):397-427. doi:10.1016/j.partic.2012.06.002
12. Lacey PMC. Developments in the theory of particle mixing. *J Appl Chem.* 1954;4(5):257-268. doi:10.1002/jctb.5010040504
13. Kramer HA (Harold A, United States. Agricultural Research Service. *Effect of Grain Velocity and Flow Rate upon the Performance of a Diverter-Type Sampler.* Washington, D.C. : U.S. Dept. of Agriculture, Agricultural Research Service; 1968. Accessed December 31, 2020. http://archive.org/details/effectofgrainvel25kram

622177 Development of Pressure Evolution Modeling for the Combustion of Distinct Metal Dust Morphologies

Abstract:

Combustible dust explosions continue to present a significant threat toward industries processing, storing, or pneumatically conveying metal dust hazards. Recent experimental investigations performed in a Siwek 20 L sphere combustion chamber allowed for the independent characterization of explosion severity as a function of variable aluminum particle morphology. Contained ignition of suspended particulates with increasing surface irregularity exhibited a distinct rise in deflagration severity – such fuel reactivity shall be modeled as a means of predicting hazardous potential for unique metal dust processing methods. Applying the shrinking particle theory with species diffusion limitations, previously reported pressure evolution outcomes were verified through the development and implementation of closed-vessel mathematical modeling based on derivation of fundamental mass and thermal balance equations. Geometric equivalence sub-models were additionally employed to offer a means of particle diameter estimation for increasingly irregular flake dust morphologies. Although the consequence prediction models demonstrated sufficient agreement with experimentally acquired data, all relevant potential shortcomings and modes of failure were examined in detail.

622114 Prediction of Solid Flow Rate during the Conveying Process Based on the Generation Mechanism of Acoustic Emission Signals

Abstract:

Prediction of solid flow rate during the conveying process based on the generation mechanism of acoustic emission signals

Peng Zhang[a], Tao Sheng[a], Hanqing Wang[b], Yao Yang[a, c, *]

[a] Zhejiang Provincial Key Laboratory of Advanced Chemical Engineering Manufacture Technology, College of Chemical and Biological Engineering, Zhejiang University, Hangzhou, 310027, P.R. China

[b] Hangzhou No.14 High School, Hangzhou 310006, P.R. China

[c] Hangzhou Global Scientific and Technological Innovation Center, Zhejiang University, Hangzhou 310027, P.R. China

* Corresponding author: Yao Yang, E-mail: yangyao306@gmail.com

It has been proved that in multiphase flow processes such as hydraulic conveying and fluidized bed, the acoustic emission signals generated by particle motions can be detected and analyzed to realize the measurement of particle size and solid mass flow rate, identification of flow regime, and fault diagnosis. However, these models are often obtained through data regression, the physical meaning of which is not clear enough. As a result, the universality of these models is bad. Therefore, this work aims to establish a predictive model of particle mass flow rate during the particle conveying process with the generation mechanism of acoustic emission signals. This method is supposed to have better university due to the clear physical meaning. The whole work is conducted as follows. Firstly, single particle-wall collision experiments are conducted to explore generation mechanism of acoustic emission signals. Results show that the intensity of the acoustic emission signals during the single particle collision is linearly related to the particle mass and particle-wall collision velocity. Then, combined with the existing homogeneous particle group-wall collision frequency model, an acoustic energy model is proposed for the homogeneous particle group-wall interaction. For the hydraulic conveying process where the particle motions were homogeneous, this model showed good prediction accuracy and universality. But for the riser with complex particle motions (particle back mixing), the model has a low accuracy since the generation of acoustic signals is so complex that it is far from the homogeneous particle group-wall collision. However, with a self-designed screened-waveguide by which the generation of acoustic emission signals is simplified into single particle-probe collision, the particle mass flow rate was also accurately predicted using the acoustic energy model.

Keywords: solid flow rate, particle-wall collision, acoustic emission detection, hydraulic conveying, riser

632723 Solar Gasification of Organic Solid Waste in an Internally-Circulated Fluidized Bed: Experiment and Modelling

Abstract:

Solar gasification is a promising and efficient approach to convert organic solid waste to high-quality syngas due to solar energy stored. Previous research outcomes of internally circulating fluidized beds have demonstrated that circulating particles inside the reaction bed is able to significantly enhance heat and mass transfer and to mitigate surface hot spots caused by non-uniform concentrated solar fluxes. In this work, a clapboard-type internally-circulating fluidized bed with an indirectly irradiated structure is experimentally studied via Singapore's first 28-kW_e high-flux solar simulator. Three different kinds of organic solid waste e.g. charcoal, chicken manure, and horticultural waste are tested under CO_2 gasification. Different experimental conditions are considered in experiment: (i) the effect of different radiated power inputs on heating rates, and (ii) the effects of the CO_2/C ratio and the bed mass on the solar upgrade factor and solar conversion efficiency. Since solar gasification operates in high-temperature solar reactors, it is difficult to adopt conventional techniques e.g. magnetic resonance imaging, particle image velocimetry, electrical capacitance tomography, etc. to investigate the fluid dynamic characteristics. Considering non-uniform solar fluxes obtained by optical calibration, a 3D numerical model is developed to indicate granular flow transport and heat transfer behaviors in solar gasification processes. This work contributes a holistic research outcome to the indirectly irradiated solar gasification of organic solid waste.

625043 Mixture Homogeneity Measurements in a Vertical Bladed Mixer Using Tracer Particles

Abstract:

Vertically bladed mixers are widely used in various solids-handling industries such as cosmetics and food and pharmaceutical manufacturing. While regulatory bodies such as the Food and Drug Administration (FDA) have strongly encouraged inline and online Process Analysis Tools (PATs) [1], traditional mixture homogeneity measurements are performed only at specific stages in the mixing process. Additionally, these methods may suffer from drawbacks due to sample extraction requirements (invasive methods) or measurement restrictions to free surfaces (noninvasive methods). In this study, mixture homogeneity is continuously measured in a bench scale vertically bladed mixer fitted with two 45° blades. Wood particles are mixed with plastic particles of varying densities, and a single wood particle is tagged with a stainless steel insert to allow X-ray radiographic particle tracking in a projected X-ray image. The tagged particle's projected location in the mixer vessel is tracked over a sufficient time period to provide the local particle distribution of all similar particles.

Mixture homogeneity values are quantified by developing a standard deviation based mixing index, a grid-independent location based mixing index [2], and the Gini coefficient [3], utilizing the tracer particle location distribution. Mixing values thus obtained are calculated continuously and demonstrate the potential for real time mixing measurements using only a few tracer particles. The effect of particle density ratio on mixing is reported and the relative differences in the sensitivity of the mixing indices to changes in mixture homogeneity are compared.

References

[1] Guidance for industry, PAT-A framework for innovative pharmaceutical development, manufacturing and quality assurance, Food and Drug Administration, Rockville, MD, 2004.

[2] M.S. Siraj, S. Radl, B.J. Glasser, J.G. Khinast, Effect of blade angle and particle size on powder mixing performance in a rectangular box, Powder Technol. 211 (2011) 100–113.

[3] A. Sen, J. Foster, On Economic Inequality, Clarendon Press Oxford, 1997.

632273 The Effect of Process and Design Parameters on Granule Content Uniformity in High Shear Granulation

Abstract:

In pharmaceutical products, uniform content of the active pharmaceutical ingredient (API) in the finished dosage form is a critical quality attribute (CQA). For high potent and low dosage drugs, the non-uniform distribution of API in the finished product can be extremely dangerous to the patient [1]. Regulatory agencies around the world including the food and drug administration of United States require the content uniformity test to performed as one of the measures to ensure quality of the final product [2]. For solid oral dosage forms such as tablets and capsules, the uniform distribution of API in the final product can be impacted by multiple unit operation such as mixing, granulation, blending, storage and tableting or capsule filling to name a few. Non uniform distribution of the API across the granule size during granulation process increases the chances of non-uniform final product due to size segregation which can occur while feeding the granules into tablet press or capsule filling machine [1]. High shear wet granulation process is often employed in pharmaceutical industry to convert the powder mix into larger granules to improve the flow property, bulk density, and to improve the homogeneity of the formulation ingredient [3,4]. Granulation also impacts the microstructure of the granules like porosity, and composition of granules across different size class which in turn affect the dissolution characteristics of the final product [4].

High shear wet granulation process steps includes a) addition and dry mixing of the raw materials using impeller in a granulation bowl b) addition of a liquid binder (binder solution or solvent) c) wet massing at higher impeller and chopper speed based on the product [5, 6]. Research studies show that the non-uniform distribution of API in granules can arise from any of these steps and can be affected by the process, formulation and design parameters of granulation [1, 3]. Particle size, material density, and the mixing conditions such as impeller speed and mixing time can contribute towards non uniform distribution of the material particles in dry mixing stage. Difference in the surface wettability of formulation component can also lead to non-uniform distribution of API by favoring or suppressing the interaction of particles with the binder droplet during the binder addition phase[1]. Coalescence, breakage, and consolidation during wet massing phase influence the re-distribution of individual components across the granule size classes [1]. The study performed to understand the origin of non-uniform distribution of material during wet granulation process by S. Oka. et al[1] shows that a sub-potent top layer at the end of dry mixing stage, during the wet granulation of a bi-component formulation, is one reason. The study also concluded that API concentration and wet massing time had the most significant impact on the uniform distribution of API in the granules.[1].

To design robust manufacturing process and minimize the potential for content non-uniformity in the granules used for producing tablets or capsules, it is important to understand the physics behind the effect of operational parameters at each processing step of wet granulation. In order to gain such an understanding, it is important to alleviate the effect of material properties of the powder mix. In this study, the fundamental process parameters of high shear wet granulation process affecting the content non uniformity of the output granules are being assessed. These parameters include loading pattern of dry powder mix, dry mixing time and impeller speed. The powder mix being used in this study represents the low dosage product. Alleviation of the impact of material property was achieved by using colored and non-colored MCC as a binary powder mix. One part of this binary powder system is colored using aqueous dye (Nigrosin), while the other part is used in its original form (white color). Such a binary powder mix eliminates the impact of primary powder particle size, shape, and wettability difference. The study focuses on assessing the impact of different loading patterns of powder such as top-bottom and side-by-side loading of formulation components in the granulator bowl, dry mixing time and impeller speed on the distribution of two components in the output granules which in turn affect the content uniformity of the granules. The homogeneity of the powder blend before the liquid addition phase is important to get the uniform output granules. The mixing studies shows that the mixing efficiency in the axial direction (top to bottom) is much smaller than the mixing in radial direction (Side to side) [7]. Thus, the loading pattern of side by side should exhibit more homogenous bed and granules with more uniform content than top bottom loading along with the change in dry mixing time and impeller speed.

Ref:

1. Oka, S., et al., *Analysis of the origins of content non-uniformity in high shear wet granulation.* International Journal of Pharmaceutics, 2017. **528**: p. 578-585.
2. Pawar, P., et al., *A "Large-N" Content Uniformity Process Analytical Technology (PAT) Method for Phenytoin Sodium Tablets.* Journal of Pharmaceutical Sciences, 2019. **108**:p. 494 -505.
3. Koide, T., et al., *Detection of component segregation in granules manufactured by high shear granulation with over-granulation conditions using near-infrared chemical imaging.* International Journal of Pharmaceutics, 2013. **441:** 135–145.
4. Oka, S., et al., *The effects of improper mixing and preferential wetting of active and excipient ingredients on content uniformity in high shear wet granulation.* Powder Technology, 2015. **278**: p. 266-277.
5. Belohlav, Z., et al., *Effect of drug active substance particles on wet granulation process.* Chemical Engineering Research and Design, 2007. Volume 85, Issue 7: p. 974-980.
6. Liu, B., et al., *A review of high shear wet granulation for better process understanding, control and product development.* Powder Technology, 2021. **381:** 204-223.
7. Gao, W., et al., *Discrete element analysis of the particle mixing performance in a ribbon mixer with a double U-shaped vessel.* Granular Matter, 2019. 21:12.

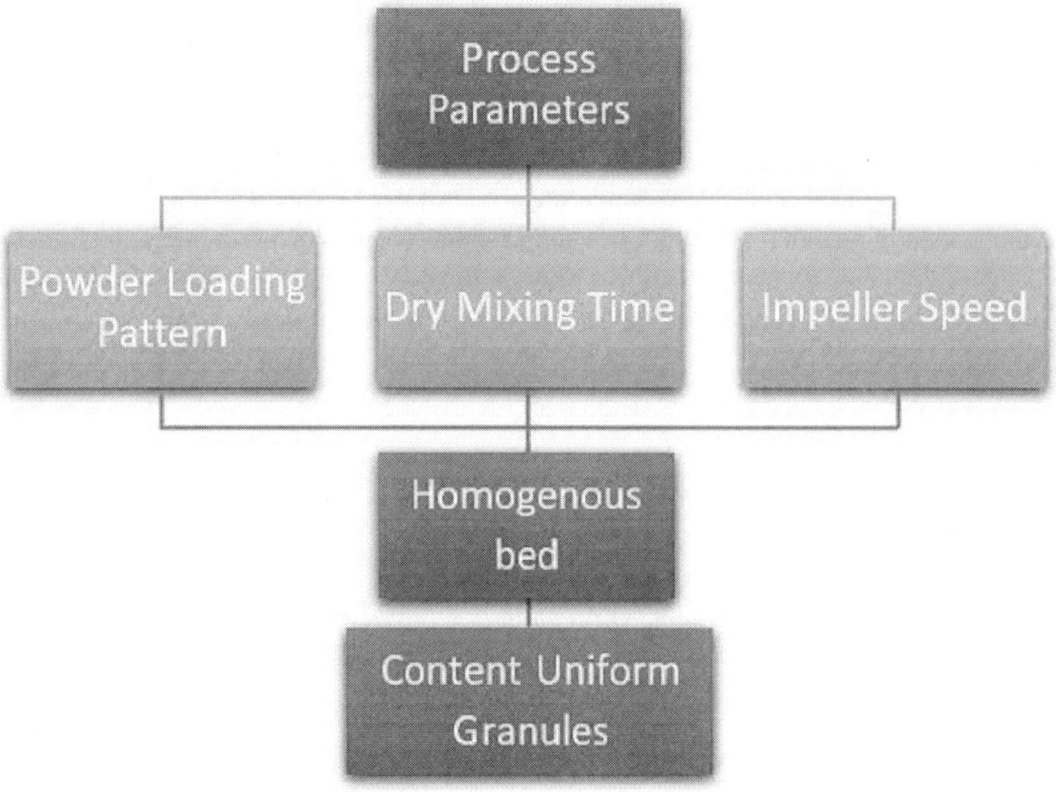
Process Parameters
Powder Loading Pattern
Dry Mixing Time
Impeller Speed
Homogenous bed
Content Uniform Granules

629482 DEM Study of a Vibrational Powder Transport System

Abstract:

Vibrational transport systems are used in the pharmaceutical industry to transport the powder from one unit operation to the next downstream one. The vibration source can be attached to a transport tube or some other powder containing vessel. The important parameters used to control the vibrational powder transport are the amplitude and the frequency, whose range is dictated by the controlling unit. Challenges may arise when transporting powders that have poor flowability, high cohesivity and adhesion levels.

The vibration system used in this work is shown in the Fig. 1. The powder is introduced from the top side in the filling hopper. The 90° elbow and the transport tube, which has a diameter reduction in one place, fill the downstream unit with the powder. The vibration source is located below the diameter reduction on the transport tube, and the whole system oscillates horizontally. The difficulties in this system include particle agglomeration, system clogging and flow blockage in the hopper and transport tube, especially because the transported material has poor flowability and is highly cohesive.

During experimental treats, it was determined that the operating frequency affects the flowability of the powder. However, to resolve the clogging and flow blocking problems using experiments is challenging.

Therefore, the Discrete Element Method (DEM) was used to model the process, accounting for millions of particles. With this approach it is possible to investigate various process conditions and equipment designs in short period of time [1], [2]. The process was closely replicated in the DEM model, with the same operating conditions and equipment design. The particles were calibrated prior to the investigation (DEM reproduction of the FT4 and compression tests), in order to closely represent the behavior of powder used in the experiments. The particle size was scaled in comparison to the original ones, keeping the distribution width as close to reality as possible.

The DEM results identified an interesting mechanism, i.e., the segregation of the smallest particles in the fill hopper, along the side walls of the system and just before the tube reduction. The investigation of the particle stress distribution in the system showed a peak of the normal and shear stress at the position of the segregated particles. The fact that these positions overlap led to the conclusion that the system clogging is caused by the small particles agglomerating when exposed to high stress. The investigation of the operating frequency showed that reducing the oscillation frequency disrupts the small particle ensembles, and works in favor of the powder flowability. This was also confirmed by experiments.

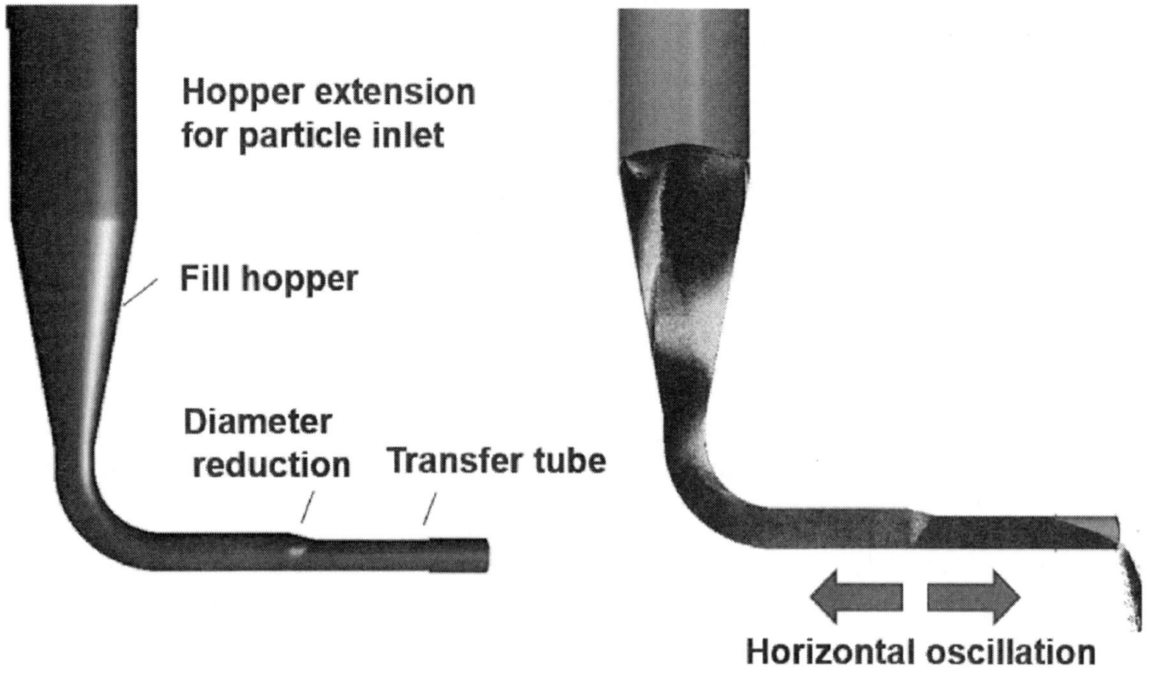
Hopper extension
for particle inlet
Fill hopper
Diameter
reduction
Transfer tube
Horizontal oscillation

629284 Rheological Investigations of Volcaniclastic Debris Flows

Abstract:

Volcaniclastic debris flows, commonly known as lahars, are hazardous natural phenomena that occur when a mixture of fine and coarse sediments and water flow down slopes after intense and/or prolonged rainfall in response to gravitational attraction. They can be generated both during or soon after explosive eruptions, but they can occur also tens or hundreds of years later.

This is the case of the area surrounding the Somma-Vesuvius and the Phlegrean Fields volcanoes in Campania region (southern Italy), where the slopes of the volcanic edifices and the Apennine reliefs are mantled by loose pyroclastic fallout and current deposits, that can be easily remobilised. Previous studies pointed out how these phenomena highly impacted on ancient settlements and how they can affect population and infrastructures during recent times, too.

In the last five years more than 500 sites among archaeological excavations, stratigraphic trenches, drill cores and outcrops have been reviewed and analysed, paying particular attention to the debris flow deposits related to the sub-Plinian AD 472 eruption of Somma-Vesuvius, characterised by a wide areal distribution. Several samples have been collected and analysed in order to characterise their grain size distribution.

The aim of this work is to acquire new knowledge in order to implement the numerical models generally used for simulating debris flows, and to compare them with the new ones. Moreover, the information acquired will be useful to assess risk and prepare mitigation plans.

For this purpose, the FT4 powder rheometer developed by Freeman Technology Ltd. has been used on several samples with the diameter of the particles less than or equal to 0.710 mm.

The shear tests have been performed with different initial consolidation stresses in order to obtain physical parameters such as the angle of internal friction, that gives an idea of the viscosity of the sample, and the flow function coefficient, that gives information on the flow behaviour. Compressibility tests have been carried out to measure how the density changes as a function of the applied normal stress. Finally, the wall friction properties have been evaluated using the wall friction test to understand how easily a powder flows against the material with which it is in contact. Sandpaper with different grits have been used to simulate different rugosities.

Preliminary results show that the angle of internal friction does not depend on the initial consolidation stresses and on the grainsize fractions. On the contrary, the flowability and the compressibility of the powders are strongly affected by the grainsize distribution and the initial consolidation stresses. Finally, the wall friction angle is affected by both grainsize distribution and the roughness of the wall. Further experiments have to be carried out to obtain additional information.

To verify the interaction between sediment and water, and so to evaluate the effect of the water content on the rheological properties of the samples, a Bohlin rheometer is employed. Furthermore, in collaboration with the British Geological Survey large-scale experiments are conducted at the University of Saint-Luis Potosí (Mexico), and the same samples used are studied to link the rheological measurements to the dynamics observed in the large-scale experiments.

622707 Segregation of Non-Spherical Particles in Vibrated Packed Bed Mixers

Abstract:

The effect of the shape of particles on mixing and segregation is studied in a vibrated packed bed mixer. Binary mixtures (unequal sized) of four types of non-spherical glass particles (prolate, oblate, rectangular, and elongated needle) with different aspect ratios are used alongside spherical particles. The overarching goal is to explore the effect of non-sphericity of granules (particles > 2 mm) used in the pharmaceutical and food industries. The total amount of energy added to the binary mixture of particles through vibrations depends on two independent operating parameters: frequency and amplitude of vibrations that are combined as a single feature defined by vibration strength. The results show the existence of a critical vibration strength below which there is little mixing. In addition, the axial segregation index value (SI) is also significantly correlated to the random monodisperse packing fraction of coarse particles. The axial segregation index follows the decreasing trend: Cube > sphere > oblate > prolate > elongated needle. Similarly, increasing the angularity of fine particles, the axial segregation is reduced. Furthermore, binary particle mixing is also affected by particle size ratio, the varying fine mass fraction, and the column's geometry. When the particle size ratio (between fine and coarse particles) it leads to reduction in segregation. The initial proportion of fines added to the system also plays an important role in determining the extent of segregation. As the proportion of the fines decreases in the initial fill - segregation decreases. In addition, the column's geometry causes a decrease in segregation when it's changed from cylindrical to rectangular.

Future work would involve using better tools like radioactive particle tracking and high-speed camera studies to make even more accurate predictions.

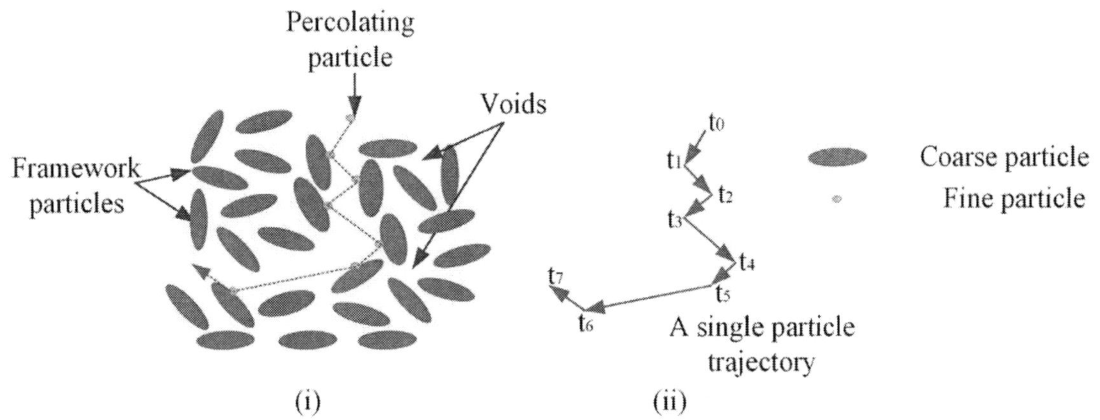

(a) Collision -Mobility-Framework for non-spherical system

Particle shape segregation in a vibrated packed bed mixer

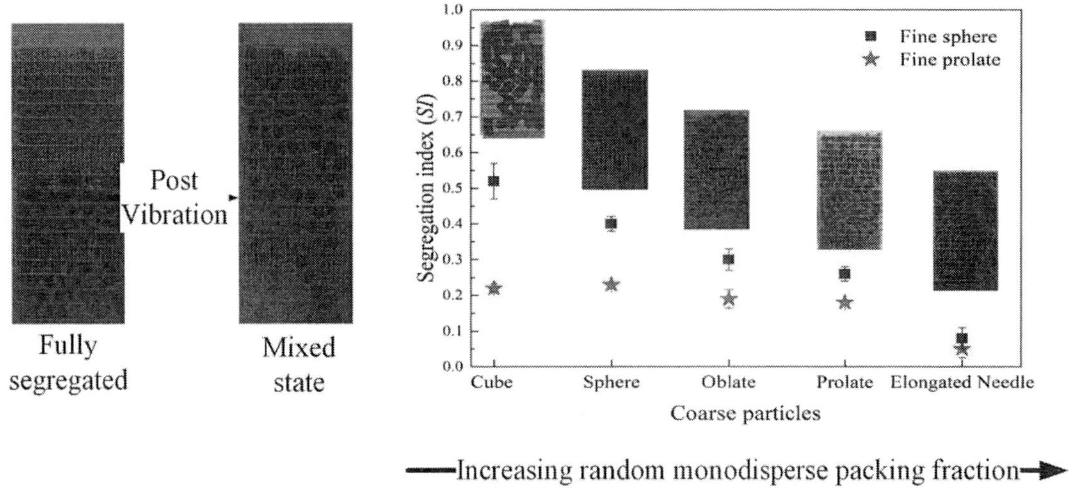

(b) Comparison of segregation index for different coarse particle shape

625473 Flow Simulations of Size Disperse Granular Particles with Realistic Contact Models

Abstract:

Accurate and general constitutive models of granular material flow would aid experimental flow characterization measurements, but yet remain elusive. Flow field, size and shape dispersity, roughness and contact mechanics change the behavior of flowing granular material and challenge existing constitutive models. Particularly, granular materials in nature and industry are rarely monodisperse and often exhibit complex distribution in particle sizes. To support the development of transferable constitutive models, we perform particle-based stress-controlled discrete element modeling simulations of particles with varying degrees of size bidispersity. Simultaneously, we also extend the simple, yet widely-used, spring-dashpot contact mechanics model by including constraints on other modes of motion. The shear stress-to-pressure ratio , density and coordination number are measured. It has been previously shown that, with slight modifications, the constitutive rheological model for monodisperse particles is applicable to granular flows with moderately polydisperse particles. Here we assess the validity of the model for extreme size bidispersity — up to 40:1 size ratio. Using our stress-controlled simulation method, we also examine flow anisotropy in such large size disperse granular flows, which is assumed to be isotropic in the standard model. Lastly, we find that the critical shear stress and flow anisotropy qualitatively changes with increasing strain rate and friction.

Sandia National Laboratories is a multimission laboratory managed and operated by National Technology &Engineering Solutions of Sandia, LLC, a wholly owned subsidiary of Honeywell International Inc., for the U.S. Department of Energy's National Nuclear Security Administration under contract DE-NA0003525. SAND2021-3774 C

622359 Calibration of Particle Properties for Multilevel/Adaptive Coarse Grained DEM Simulations

Abstract:

Calibration of particle properties for DEM is an important area with many open questions and unresolved issues. One of the ways to obtain the particle properties for DEM is bulk calibration[1]. In this method, DEM simulation results are matched with a set of experimental observations such as dynamic and static angle of repose, hopper discharge rate etc. to obtain the best fit values of the parameters. However, most of the studies are aimed at calibrating the actual particle properties, and literature on property calibration of coarse-grained particles is sparse. In this work we will address this research gap by designing a calibration methodology for the recently developed multilevel/adaptive coarse-graining technique [2]. In this technique the coarse-graining ratio of a particle changes dynamically during its course of motion through the equipment, unlike a more commonly used constant coarsegraining ratio.

In this presentation, we will first discuss the relation between the microparameters of the resolved particles and that of the coarse-grained particles. We will show a study that explores various force scaling, like that discussed by Sakai et al. [3], and demonstrate their effect on the final model outcome. The scaling for other microparameters such as particle density, particle to wall friction etc. will also be discussed. Finally, a systematic method will be demonstrated which can be used for calibration of the particle properties at various zones of an adaptive coarse-grained simulation. A tablet die feeder and a screw feeder will be used as two examples

References:

1. C. Coetzee, "Calibration of the discrete element method,", Powder Technology, vol. 310, pp. 104–142, 2017
2. (320b) A Multi-Level Coarse Graining Technique for Discrete Element Method (DEM) Simulation. Jayanta Chakraborty, Tarun De, Jitendra Kumar, Salvador García-Muñoz, Maitraye Sen. AIChE Annual Meeting, 2020.
3. M. Sakai and S. Koshizuka, "Large-scale discrete element modeling in pneumatic conveying," Chemical Engineering Science, vol. 64, no. 3, pp. 533–539, 2009

630549 Eulerian-Eulerian Modeling of Segregation and Dispersion in Multi-Component Fluidized Beds: Verification, Validation and Limitations

Abstract:

In many industrial-scale fluidized-bed reactors, particle mixing and segregation play a vital role in determining the reactor performance. Detailed information about the phase voidage distribution throughout the bed at different operating conditions is essential for design, scale-up, and optimization. Several experimental and theoretical studies are published in literature focusing on solid-liquid fluidized beds (SLFB) consisting of mono and bi-dispersed particle systems. However, very few studies in the literature emphasize the fluidization of systems involving more than two types of particles. Such kinds of systems are of high industrial importance in mineral and chemical process industries. Therefore, it is desirable to study the coupling between superficial liquid velocity and multi-size (and multi-density) particle systems consisting of more than two types of particles.

Multi-component SLFB cases exhibit several possibilities: complete segregation, partial mixing, segregation with an interface, and segregation without interface. A few experimental investigations have also shown that the layer inversion phenomena exhibited by SLFB at specific conditions also exist for ternary systems. In the present work, Eulerian-Eulerian simulations have been carried out to investigate particle density and diameter on the segregation and dispersion behavior in SLFB comprising of two to six different sizes and densities of solid phases.

In the initial part of this study, many bidisperse systems are simulated to evaluate the advantages and limitations of the commonly used boundary conditions, geometric simplifications, and empirical models used for modeling of SLFB. The limitations identified are (i) inadequacy of the fluid-solid drag models in accounting for the hindrance effect, (ii) unsuitability of solid-solid interaction force closures for the flow physics exhibited by the SLFB, and (iii) the broad spectra of time range at which different systems achieve a steady state. The identified limitations are addressed by introducing physically appropriate empirical models, boundary conditions, and running the simulations for sufficiently long-time steps.

The improved modeling approach was then used to simulate multi-component fluidized beds. The solid fractions, pressure drop, and voidage profiles obtained from the Eulerian-Eulerian simulations of multi-component systems were compared with the experimental data and theoretical model predictions. The simulations show that although Eulerian-Eulerian simulation may not be as detailed and accurate as the Discrete Element Model (DEM), they still provide precise predictions of solid axial profiles, voidage, and overall segregation, and dispersion behavior. However, the accurate predictions of the SLFB hydrodynamics completely depend upon the appropriate selection of the interfacial closure models, boundary conditions, and time range over which the simulations are run. Moreover, the computational time required to simulate a multi-component SLFB using the Eulerian-Eulerian model is significantly less (~10 times) than that required by DEM to simulate the same system. Thus, the present investigation proposes an economically viable and rational approach for CFD and process engineers to get quick estimates of the SLFB hydrodynamics.

621522 A Nano-CT Informed Polyhedral Discrete Element Modeling Approach for Flow of Complex-Shaped Granular Woody Biomass

Abstract:

The design of handling equipment for granular biomass primarily requires experiments. Discrete element models (DEM) can provide designers with detailed insight into the behavior of granular materials. However, granular biomass comprising complex-shaped particles is difficult to model with DEM. A tomography-informed DEM approach and an exhaustive assessment of the ability of the approach to predict the bulk behavior of milled pines using experimental data is presented. Nano-CT scan was conducted to obtain 3D particle surface geometries as the basis for particle shape approximation by a polyhedral model and a sphero-polyhedral model. These models were applied in the simulation of a compressibility test. Our parametric study showed that particle Young's modulus and restitution coefficient are the two main properties influencing the simulated bulk behavior of the DEM particles and that the level of approximating DEM particle shape for real particles is critical for accurately replicating the bulk behavior of milled pines. The polyhedral model demonstrated better suitability than its sphero-polyhedral counterpart for modeling pine particles. The polyhedral model calibrated in a compressibility test was then applied in the simulations of a friction test without additional parameter tuning. Interlocking was dominant in the shear of bulk pine particles. Remarkably, the polyhedral model predicted the frictional behavior of the pine particles when compared to the experimental results. The limitations of the model, as well as possible ways for enhancement, are discussed. This work provided novel insights into the suitability of complex-shaped DEM models for granular woody biomass.

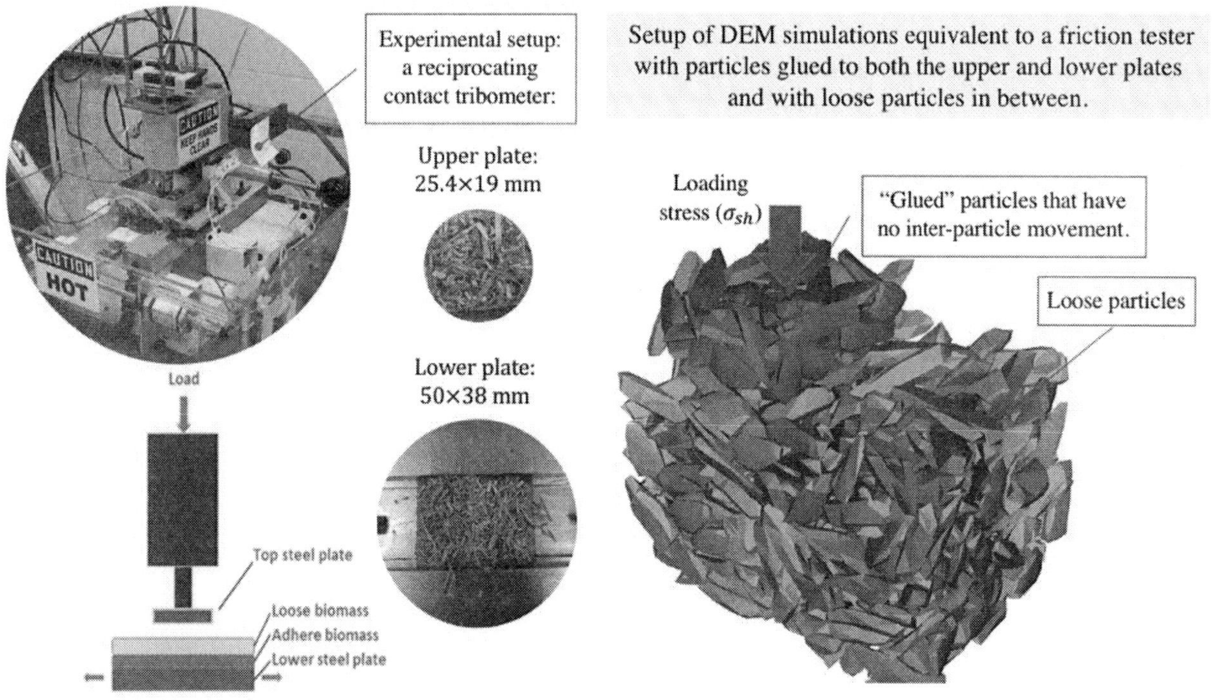

621550 A Class of Generalized Strain-Hardening Discrete Element Method (DEM): Theory, Ligghts Open-Source Implementation, and Applications for Granular Biomass Flow

Abstract:

In this work we present a class of generalized strain-hardening contact force-displacement models for the discrete element method (DEM). The development of these contact models is inspired by the experimentally measured bulk stress-strain behaviors of fractured granular woody biomass feedstocks such as chipped pine. Granular biomass in general exhibits complex and irregular particle shapes and relatively low stiffness compared with hard, regular-shaped bulk solids such as pharmaceutical pellets and agricultural products. These features of granular biomass result in its unique bulk strain-hardening mechanical characteristics. Although the polyhedral DEM for arbitrary particle shapes can approximate the complex-shaped woody biomass particles with fine geometric details, the computational cost required by the polyhedral DEM can be prohibitive for simulating even a few thousands of particles, if each polyhedral particle comprises of a few hundreds of surface triangular elements for high-quality shape approximation. To allow for the DEM to simulate large-scale transport and feeding operations of granular biomass in the biorefineries, we envisioned a coarse-grained (CG) DEM approach by using only spherical particles as the representative elementary volumes (REV) of bulk granular biomass. To compensate for the loss of accuracy in the particle shapes, we developed a class of strain-hardening force-displacement models for particle-particle contact. The formulation of an unconditionally stabilized viscous damping force associated with each of the contact models is also derived. Since these models are intended for general adoption of the DEM application users rather than biomass flow applications alone, the mathematical formulations in these models are kept in the generalized polynomial and/or exponential functions. To allow for the best possible access to these models on all general computing platforms (PC, macOS, Linux, HPC, etc.), we implemented these models based on the LIGGGHTS open-source DEM particle simulation software and released the LIGGGHTS-INL package (

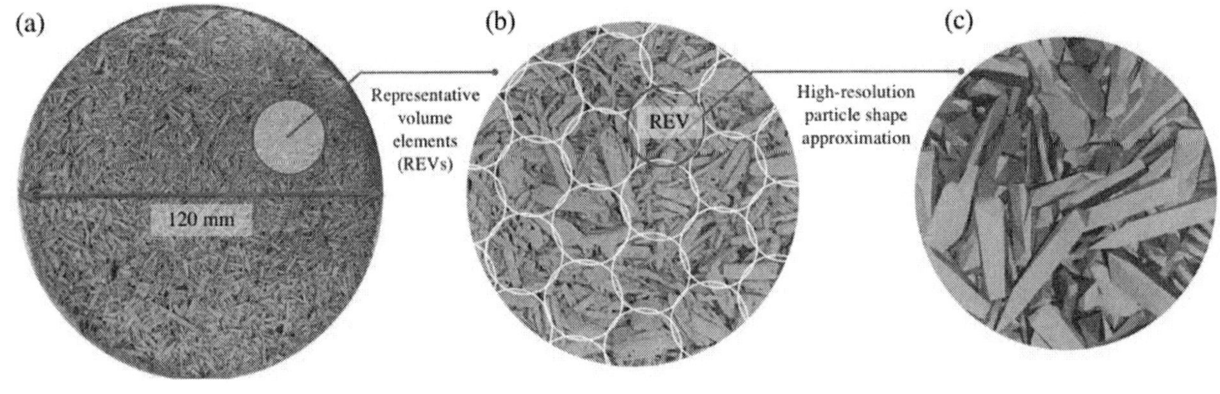

629274 Enhanced Flowability of Dense Bi-Disperse Granular Avalanches: The Role of Fine Grains

Abstract:

Dense granular flow on inclined surfaces is encountered in engineering applications as well as in geophysical flows (e.g., pyroclastic flows, landslides and avalanches) [1]. Understanding the flowability of dense granular flow is paramount for both industrial optimisation and mitigation and prevention of geo-hazards. Polydispersity is an inherent attribute of both natural and manmade granular matters. A large number studies of dense granular flow deals with monodisperse or slightly polydisperse granular media [e.g., 2,3], despite granular polydispersity has been proved to be a critical factor in dense granular flow behaviour [4]. In this work, we study the flowability of dense bi-disperse granular avalanches using 3D Discrete Element Method (DEM). The objectives are to reveal the flowing features of the granular system and explore the physical mechanisms underlying the observed physical behaviour.

To pursue the objectives of this work, a flume model, which is composed of two main parts, namely an acceleration zone (inclined plane) and deposition zone (horizontal plane), is employed. Two kinds of spherical cohesionless particles, differing only in size (size ratio of 2.5), are used. We keep the total mass of particles constant and vary the mass proportion of large particles. The initial granular assembly is generated by randomly pouring bi-disperse particles into a reservoir under gravity to preserve a well-mixed initial state. The contact forces between particles are simulated using a linear spring-dashpot model. Both sliding and rolling friction are considered. The coefficient of restitution for all types of collisions is kept the same by appropriately selecting the damping coefficient. The DEM simulations are conducted using the open-source code LAMMPS (Largescale Atomic/Molecular Massively Parallel Simulator) [5].

Similar flowability, or runout, is observed for monodisperse granular assemblies, which is independent of granular size. However, bi-disperse granular avalanches present enhanced flowability and maximum flowability is observed at a critical ratio of small and particles. Moreover, reverse grading, in which small particles are found at the bottom and at the rear zone of the control volume, is captured. Despite the rotational energy is small in comparison with translational energy, the rotation speed of small particle is greatly increased in dense bi-disperse granular flow. To explore possible mechanisms responsible for such a feature, we studied the interactions between large and small particles in term of the restitution coefficient, sliding friction and rolling friction. We found that the rolling friction between large and small particles plays a critical role in dense bi-disperse granular flow. Specifically, the rotation of small particles can efficiently lubricate bi-disperse the granular assembly to increase its flowability. Although more simulations will be needed using large polydisperse granular media, this work provides very useful insights into understanding the effects of polydispersity on granular flowability.

Reference

[1] Goujon, C, B Dalloz-Dubrujeaud, N Thomas. Bidisperse granular avalanches on inclined planes: A rich variety of behaviors. The European Physical Journal E. 2007;23(2):199-215.

[2] Chialvo, S, J Sun, S Sundaresan. Bridging the rheology of granular flows in three regimes. Physical review E. 2012;85(2):021305.

[3] Zhu, C, Y Huang, J Sun. Solid-like and liquid-like granular flows on inclined surfaces under vibration–Implications for earthquake-induced landslides. Computers and Geotechnics. 2020; 123:103598.

[4] Remy, B, JG Khinast, BJ Glasser. Polydisperse granular flows in a bladed mixer: experiments and simulations of cohesionless spheres. Chemical Engineering Science. 2011;66(9):1811-1824.

[5] Plimpton, S. Fast parallel algorithms for short-range molecular dynamics. Journal of Computational Physics. 1995;117(1):1–19.

627169 DEM Simulation of Bin Blending Considering Different Scales

Abstract:

Bin blending is a crucial step in the solid dosage form manufacturing in pharmaceutical industries. One of the main challenges in this process is reaching the desired content uniformity while avoiding over-blending. Over-blending may increase the powder cohesiveness, especially in cases where the amount of lubricant is low. This may disturb the subsequent processes i.e., tablet press or capsule filling and preferred to be avoided. Also, over-blending may lead to over-lubrication, reducing bioavailability.

Several studies were performed in this field with a main focus on content uniformity, whereas over-blending drew less attention. In this study, experiments and discrete element method (DEM) simulations are implemented to study the blending process with a focus on both content uniformity and over blending effect. According to our experiments, powder cohesiveness is increasing during the process. However, the trend of this increase is not the same at different scales. At the smallest scale (lab scale), one hour of blending leads to have a cohesive powder. In contrast, in the production scale, around 10 minutes blending brings about a cohesive product which challenges the blending time. In order to understand the reasons for this behaviour and to avoid them, 4 scales of a bin blenders (from 2.5 litre to 100 litre) considering various operational conditions are simulated in this work. The effect of rotational speed (rpm), loading mass percentage, and the powder cohesiveness is investigated by means of particles velocity, travelling distance, shear stress, and blending time. Due to the computational limits of simulating 2.5 kg to 45 kg of powder with real particle size distribution, particles are scaled considering DEM contact force model parameter calibration. A total 35 cases with the number of particles in the range of 4 to 60 million are modelled. For the post-processing, studied parameters, i.e., travelling distance, shear stress, and blending time are extrapolated by applying the relay race method. Moreover, particles velocity is calculated based on a rotating cylindrical coordinate system considering blender rotational axis as the cylinder axis.

Results can be divided into three main sections: 1. Detailed study of the influence of operational condition on the studied parameters. 2. The correlation between parameters crossing different scales. 3. The relation between the powder cohesiveness gained from experiments and studied parameters from DEM simulation. In the first section, it is observed that by increasing loading mass the velocity and travel distance is decreased. This causes longer blending time compare to less loading mass in the system. Moreover, particles are exposed to higher shear stress when the bin is loaded with more mass. Rotational speed investigation reveals that particles at the surface layers gain higher speed by increasing the RPM. In contrary, shear stress on particles is not affected by rotational speed. Moreover, results of the simulations with a more cohesive powder show dramatic increase in shear stress. Second section brings the geometry size effect on the studied parameters. According to the results, blending time is mostly influenced by the loading mass percentage in the system in different scales. However, the shear stress is higher in the larger blenders. Particle velocity and distance travel are also increased by the bin size. Finally, in the third section, powder flow function coefficients (FFC) gained from blended powder in various operational condition are correlated with the shear stress from the respective simulations. According to the results, powder FFC decreases with the particles exposed to shear. In other words, conditions, such as higher loading and longer blending time, are making powder more cohesive. The cut-off time of the acceptable FFC value can be selected from the trend and compared to blending time. If content uniformity is not satisfied through this time, the loading mass inside the bin should be reduced to reach the content uniformity as well as avoid cohesive product.

the loading mass inside the bin should be reduced to reach the content uniformity as well as avoid cohesive product.

623247 New Developments in Positron Emission Particle Tracking and Their Applications in Food, Pharmaceuticals and Plastic Recycling

Abstract:

Positron emission particle tracking (PEPT) is a non-invasive imaging technique capable of providing detailed, high-resolution, three-dimensional data even from the interior of large, dense, optically opaque systems (1). Though the technique has existed for more than 3 decades, recent years have seen an upsurge in new developments and improved capabilities. These include: the development of novel, machine-learning based algorithms (2) capable of, for the first time, of resolving particle collisions; a detailed and accurate simulation model of the Birmingham positron camera used in hundreds of previous experiments (3); the development and inauguration of a new "SuperPEPT" detector system offering micron-scale spatial resolution and sub-millisecond temporal resolution; the development of a first set of benchmarking tests allowing PEPT algorithms developed around the world to be rigorously tested and quantitatively compared; and the hybridisation of PEPT with DEM simulations and genetic algorithms to autonomously characterise particles and powders. This talk will provide a flying overview of these new developments, and the manners in which they have been applied to a diverse range of chemical engineering problems from refining the measurement of just-suspended speeds in stirred-tank systems (4) to the imaging and optimisation of mixing processes within novel fluidised-bed-based waste-plastic recycling methodologies (5, 6).

REFERENCES

1. Windows-Yule C, Seville J, Ingram A, Parker D. Positron Emission Particle Tracking of Granular Flows. Annual Review of Chemical and Biomolecular Engineering. 2020;11.
2. Nicuşan A, Windows-Yule C. Positron emission particle tracking using machine learning. Review of Scientific Instruments. 2020;91(1):013329.
3. Herald M, Wheldon T, Windows-Yule C. Monte Carlo model validation of a detector system used for Positron Emission Particle Tracking. Nuclear Instruments and Methods in Physics Research Section A: Accelerators, Spectrometers, Detectors and Associated Equipment. 2021;993:165073.
4. Windows-Yule CR, Hart-Villamil R, Ridout T, Kokalova T, Nogueira-Filho JC. Positron Emission Particle Tracking for Liquid-Solid Mixing in Stirred Tanks. Chemical Engineering & Technology. 2020;43(10):1939-50.
5. Windows-Yule C, Moore A, Wellard C, Werner D, Parker D, Seville J. Particle distributions in binary gas-fluidised beds: Shape matters–But not much. Chemical Engineering Science. 2020;216:115440.
6. Windows-Yule C, Gibson S, Werner D, Parker D, Kokalova T, Seville J. Effect of distributor design on particle distribution in a binary fluidised bed. Powder Technology. 2020;367:1-9.

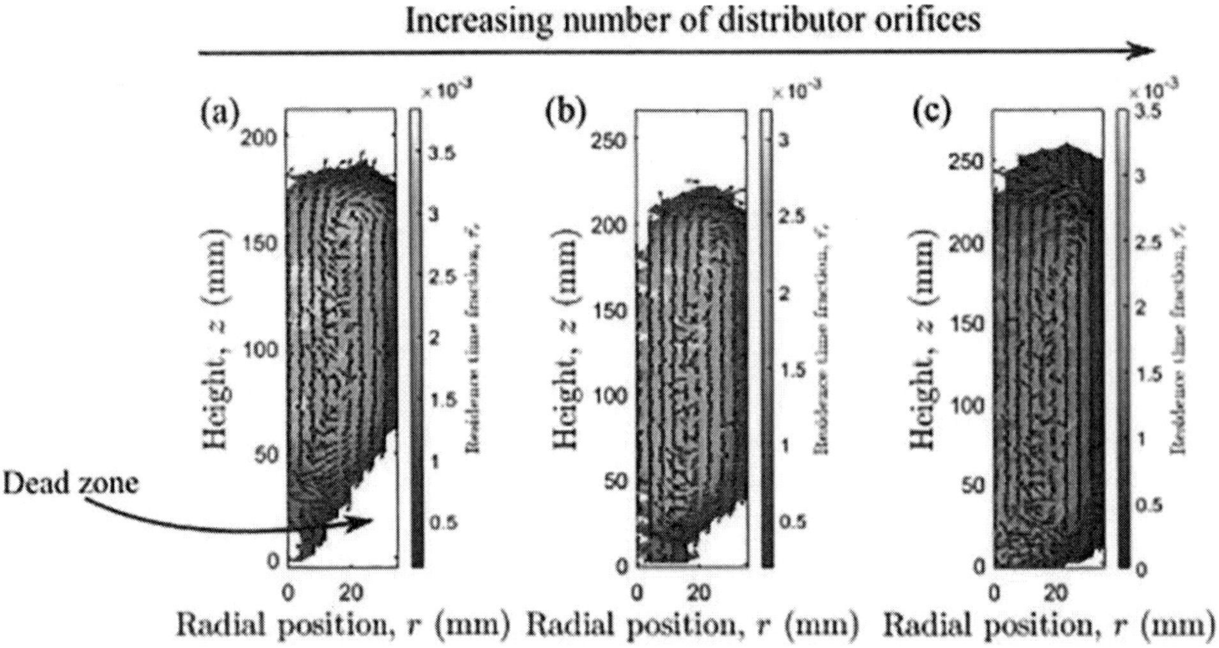
Increasing number of distributor orifices
(a)
(b)
(c)
Dead zone
Height, z (mm)
Radial position, r (mm)
Residence time fraction
× 10⁻³

626403 Calibration of DEM Parameters for Cohesive Powders Using an Annular Shear Cell

Abstract:

Discrete Element Method (DEM) modeling of particulate flows is becoming increasingly common for designing, troubleshooting, and optimizing industrial processes. Creating an effective DEM model requires an efficient methodology to calibrate material properties. In this work, a DEM parameter calibration strategy for cohesive powders is proposed, specifically for particle-particle sliding friction and particle-particle cohesive surface energy density. This method involves using a single, common bulk powder flow testing device, the shear cell, to obtain predictions of incipient yield loci and critical state loci for a range of DEM parameters with simulations planned using design of experiments. Dimensional analysis is used to further reduce this parameter space and increase the generality of the results. The simulation results are analyzed to find the statistically significant parameters, and any major interaction effects between two parameters which are then used to create a predictive model using regression. These predictions are compared with the experimental values to determine the appropriate DEM particle parameter values. The effects of particle size distribution, shear cell pre-consolidation stress, particle-particle sliding friction, particle-particle cohesive surface energy density, and particle aspect ratio on the shear cell bulk behavior as predicted by DEM are also explored. These DEM parameters, calibrated for three different pharmaceutical materials, are validated by modelling an independent system and comparing the results to those from experiments.

622662 DEM Modelling of Battery Electrode Structures from X-Ray Tomography-Structure Analysis Under Different Calendering Conditions

Abstract:

The calendering process is an essential step in the manufacture of electrodes for lithium ion batteries. Calendering strongly affects the microstructure and end-product performance, however the relationship between the various component material properties and operating conditions of the calendering process on electrode microstructure and performance is not well understood.

In this work, Discrete Element Method (DEM) was used to investigate the electrode structure evolution under different calendering conditions. The initial positions of active material (AM) particles are obtained from an uncalendered electrode microstructure characterized experimentally by X-ray tomography, and imported into DEM simulations. Simulated electrode structures under varying calendering conditions are obtained by uniaxial compression tests in DEM. The Edinburgh Elastic Plastic Adhesive (EEPA) model and bond model are used to describe the mechanical properties of AM particles and binder phase within electrodes. Detailed stress and structural evolutions at microscopic scale are further analysed. The results demonstrate that DEM provides a promising way to investigate and optimize the particle arrangements within battery electrodes during calendering.

AUTHOR INDEX

AIChE
120 Wall Street, FL 23
New York, NY 10005-4020

ISBN 978-1-7138-5699-3

9 781713 856993

CURRAN ASSOCIATES INC.
proceedings.com